Bayesian Machine Learning in Geotechnical Site Characterization

Bayesian data analysis and modeling linked with machine learning offers a new tool for handling geotechnical data. This book presents recent advancements made by the author in the area of probabilistic geotechnical site characterization.

Two types of correlation play central roles in geotechnical site characterization: cross-correlation among soil properties and spatial correlation in the underground space. The book starts with the introduction of the Bayesian notion of probability "degree of belief", showing that well-known probability axioms can be obtained by Boolean logic and the definition of plausibility function without the use of the notion "relative frequency". It then reviews probability theories and useful probability models for cross-correlation and spatial correlation. Methods for Bayesian parameter estimation and prediction are also presented, and the use of these methods is demonstrated with geotechnical site characterization examples.

Bayesian Machine Learning in Geotechnical Site Characterization suits consulting engineers and graduate students in the area.

Jianye Ching is Distinguished Professor at National Taiwan University and Convener of the Civil & Hydraulic Engineering Program of the Ministry of Science and Technology of Taiwan. He is/was Chair of ISSMGE's TC304 (risk), Chair of Geotechnical Safety Network (GEOSNet), and Managing Editor of the journal *Georisk*.

Challenges in Geotechnical and Rock Engineering

This series offers advanced level books focusing on state-of-the-art methods for handling problems across geotechnical engineering.

Chief Editor *Kok-Kwang Phoon, Singapore University of Technology and Design*

Assistant Editor *Dong-ming Zhang, Tongji University*

Evolutionary Process of a Steep Rocky Reservoir Bank in a Dynamic Mechanical Environment
Luqi Wang and Wengang Zhang

Uncertainty, Modelling, and Decision Making in Geotechnics
Edited by Kok-Kwang Phoon, Takayuki Shuku and Jianye Ching

Bayesian Machine Learning in Geotechnical Site Characterization
Jianye Ching

For more information about this series, please visit: www.routledge.com/Challenges-in-Geotechnical-and-Rock-Engineering/book-series/CGRE

Bayesian Machine Learning in Geotechnical Site Characterization

Jianye Ching

CRC Press
Taylor & Francis Group
Boca Raton London New York

CRC Press is an imprint of the
Taylor & Francis Group, an **informa** business

Front cover image: Tatyana20/Shutterstock.com

First edition published 2025
by CRC Press
2385 NW Executive Center Drive, Suite 320, Boca Raton FL 33431

and by CRC Press
4 Park Square, Milton Park, Abingdon, Oxon, OX14 4RN

CRC Press is an imprint of Taylor & Francis Group, LLC

ISBN: 978-1-032-31441-9 (hbk)
ISBN: 978-1-032-31443-3 (pbk)
ISBN: 978-1-003-30976-5 (ebk)

DOI: 10.1201/9781003309765

Typeset in Sabon
by SPi Technologies India Pvt Ltd (Straive)

Contents

Series Preface

CHALLENGES IN GEOTECHNICAL AND ROCK ENGINEERING

Geotechnical and rock engineering have made significant strides in response to different challenges and opportunities that include measuring and understanding material/structural behaviors, handling special environmental conditions, performing complex numerical simulations, design, construction, and circular role for low-carbon materials, novel structures, green construction and operation, life cycle and risk/reliability informed management, data-driven algorithms and AI-based decision-making, autonomous systems, digital twin and smart infrastructure, resilience engineering, climate change, and sustainability, among others. These challenges are interrelated. Although some of the challenges are common to all industries, it is not meaningful to engage them in an abstract manner outside of the practice context. One example is data-centric geotechnics that take a "data first" approach to decision-making, but the data are actual "ugly" field observations (multi-source, sparse, incomplete, spatially variable, corrupted, etc.) rather than ideal abstract numbers. Data-centric geotechnics should deploy digital technologies in cognizance of the context of geotechnical and rock engineering that includes physics, empirical knowledge, experience, and engineering judgment. How decision-making in geotechnical and rock engineering can be revolutionized through human–machine collaboration is one grand challenge that epitomizes the motivation for launching this book series. This book series presents exciting emerging solutions in geotechnical and rock engineering that are expected to transform practice and meet fast-evolving environmental/societal trends in the twenty-first century. It is a timely response to the changing technological, environmental, and societal landscape presented in the Institution of Civil Engineers (ICE) State of the Nation Report on "Digital Transformation" and the American Society of Civil Engineers (ASCE) "Future World Vision: Infrastructure Reimagined" paper.

Book series editors

Kok-Kwang Phoon
Dongming Zhang

Preface

Variability and uncertainty are inevitable during the process of geotechnical site characterization. There is inherent (spatial) variability in soil properties (Lumb 1966; e.g., Vanmarcke 1977, 1983; Phoon and Kulhawy 1999a). Measurement errors may occur during laboratory or in situ tests. Transformation models (e.g., the empirical equation that converts CPT cone tip resistance to undrained shear strength) are not exact, so there is transformation uncertainty (Phoon and Kulhawy 1999b). Site investigation data are typically limited, so there is statistical uncertainty (Phoon and Kulhawy 1999a). There are two different statistical approaches for data analysis to deal with variability and uncertainty: frequentist and Bayesian approaches (Baecher and Christian 2003).

The frequentist approach considers probability as "relative frequency". In this approach, an unknown parameter is considered as a fixed constant (not random). A point estimate for the parameter can be computed based on measured data, and the confidence interval (CI) can be established as well. There are two main limitations to the frequentist approach when it is applied to geotechnical engineering. First, it usually requires a significant amount of data to construct a useful (e.g., narrow) CI. However, in geotechnical engineering, it may not be feasible to conduct many tests. For instance, it is common that less than ten undisturbed soil samples are available to determine the design soil shear strength parameter. It is also common that a single load test is conducted to determine the pile capacity parameter. Second, the significance of the 95% CI in the frequentist approach is not very useful in practice. The significance of the 95% CI is rather deep. Suppose that N test programs can be conducted to produce N sets of 95% CIs. The significance of the 95% CI is that on average, there are 95% × N sets of these CIs containing the actual parameter. If only a single test program is conducted to obtain a single 95% CI. In principle, the chance of this 95% CI containing the actual parameter is either 0 or 100% because the parameter is fixed.

The Bayesian approach (e.g., Ang and Tang 1975, 2007; Beck and Katafygiotis 1998; Baecher and Christian 2003; Yuen 2010; Gelman et al. 2013) considers probability as a "degree of belief". In this approach, an unknown parameter is considered "random" or "uncertain". A prior probability distribution is required to quantify the uncertainty in the parameter, and this prior distribution can be updated into the posterior distribution by the measured data via the Bayes rule. This Bayesian updating does not require a significant amount of data. The Bayesian confidence interval (formally called the credibility interval) can be established based on the posterior distribution. The significance of the Bayesian 95% CI is simple: it has a 95% chance of containing the actual parameter, even if only a single experiment is conducted to obtain the Bayesian CI. There are two main limitations to the Bayesian approach. First, the Bayesian approach requires the prior distribution. The prior distribution is more or less subjective in the absence of data. Second, the Bayesian approach usually requires complex computation procedures such as Markov chain Monte Carlo (MCMC) (Gilks et al. 1998) and higher computational costs.

In the context of geotechnical engineering, the limitations of the Bayesian approach are more amendable than those of the frequentist approach. For the Bayesian approach, the subjectivity in the prior may be minimized by adopting the so-called non-informative prior. The high computational cost (e.g., computation time) may be alleviated by advancements in computation power (e.g., high-performance PCs). For the frequentist approach, however, the requirement for a significant amount of data is difficult to achieve for most geotechnical engineering projects. Also, the significance of the frequentist CI may be too deep to be grasped correctly by average geotechnical engineers. In the opinion of the author of this book, the Bayesian approach is more useful than the frequentist one in terms of geotechnical applications. In one of his lectures, Prof. Gregory Baecher (Baecher 2019) stated that the "frequentist statistical methods are mostly useless in geotechnical applications".

The purpose of this book is to present recent advancements made by the author in the area of probabilistic geotechnical site characterization. Two types of correlations play central roles in geotechnical site characterization: cross-correlation among soil properties and spatial correlation in the underground space. The book starts with the introduction of the Bayesian notion of probability "degree of belief", showing that well-known probability axioms can be obtained by Boolean logic and the definition of plausibility function without the use of the notion "relative frequency". The book then reviews probability theories and useful probability models for cross-correlation and spatial correlation. Methods for Bayesian parameter estimation and prediction will be presented. Finally, the use of these methods will be demonstrated by geotechnical site characterization examples.

Jianye Ching

Chapter 1

Bayesian approach

There are two notions of probability: relative frequency and degree of belief. The frequentist approach takes the former view, whereas the Bayesian approach takes the latter. The purpose of this chapter is to show that the degree-of-belief notion yields several Bayesian probability axioms. Interestingly, these Bayesian probability axioms have the same algebraic form as their frequentist counterparts. Many of the contents of this chapter are extracted from the lecture note made by Prof. James L. Beck for the 2004 summer lecture series held at Caltech (Instructors: James L. Beck, Jianye Ching, and Ivan Au), which is based on the works done by Cox (1946, 1961). The discussion starts with the Boolean logic for propositions, followed by the definition of the "plausibility" and "probability" of a proposition. This probability is the degree of belief for a proposition, not the relative frequency, that is, it is the Bayesian probability. Finally, it is shown that the Bayesian probability must satisfy several probability axioms. A similar topic called "probability logic" has been discussed in Beck (2010).

1.1 BOOLEAN LOGIC

Let us first define a "proposition": A proposition is a statement that makes an assertion, that is, true or false. For instance, **a** = "speed of a vehicle exceeds 60 mph" is a proposition because one can argue that this statement is true or false. Not all statements are propositions, for example, daily conversations such as "good morning" are statements, but they are not propositions.

There are several basic logic operations over propositions. These operations produce new propositions based on existing proposition(s):

1. Negation of **a**: ~**a** ("not **a**")
   The "truth value" of **a**, denoted by t(**a**), is the indicator function of **a**:

$$t(\mathbf{a}) = \begin{cases} 1 & \text{if } \mathbf{a} \text{ true} \\ 0 & \text{if } \mathbf{a} \text{ false} \end{cases} \tag{1.1}$$

   The relationship between t(~**a**) and t(**a**) is shown in Table 1.1.

DOI: 10.1201/9781003309765-1

Table 1.1 Relationship between t(**a**) and t(~**a**)

t(a)	*t(~a)*
0	1
1	0

Table 1.2 Relationship between t(**a & b**) and [t(**a**), t(**b**)]

t(a)	*t(b)*	*t(a & b)*
1	1	1
1	0	0
0	1	0
0	0	0

Table 1.3 Relationship between t(**a or b**) and [t(**a**), t(**b**)]

t(a)	*t(b)*	*t(a or b)*
1	1	1
1	0	1
0	1	1
0	0	0

2. Conjunction of **a** and **b**: **a & b** ("**a** and **b**")
   The relationship between t(**a & b**) and [t(**a**), t(**b**)] is shown in Table 1.2.
3. Disjunction of **a** and **b**: **a or b** ("**a** or **b**")

The relationship between t(**a or b**) and [t(**a**), t(**b**)] is shown in Table 1.3.
Let us now define the notation of "logical equivalence", denoted by the sign "≡". It is similar to the equality sign "=" for numbers. There are several laws for logical equivalence:

$$\text{Commutative law: } (\mathbf{a} \text{ or } \mathbf{b}) \equiv (\mathbf{b} \text{ or } \mathbf{a}) \tag{1.2}$$

$$\text{Associative law: } \mathbf{a} \,\&\, (\mathbf{b} \,\&\, \mathbf{c}) \equiv (\mathbf{a} \,\&\, \mathbf{b}) \,\&\, \mathbf{c} \tag{1.3}$$

$$\text{Distributive law: } \mathbf{a} \,\&\, (\mathbf{b} \text{ or } \mathbf{c}) \equiv (\mathbf{a} \,\&\, \mathbf{b}) \text{ or } (\mathbf{a} \,\&\, \mathbf{c}) \tag{1.4}$$

$$\text{Double negation law: } \sim(\sim \mathbf{a}) \equiv \mathbf{a} \tag{1.5}$$

$$\text{De Morgan's laws: } \sim(\mathbf{a} \text{ or } \mathbf{b}) \equiv \sim\mathbf{a} \,\&\, \sim\mathbf{b}, \sim(\mathbf{a} \,\&\, \mathbf{b}) \equiv \sim\mathbf{a} \text{ or } \sim\mathbf{b} \tag{1.6}$$

Two propositions are logically equivalent if their truth values for all possible true–false combinations are all equal. Let us take the second De Morgan's

Table 1.4 Truth values for the second De Morgan's law

t(**a**)	*t*(**b**)	*t*(**a** & **b**)	*t*(~(**a** & **b**))	*t*(~**a**)	*t*(~**b**)	*t*(~**a** or ~**b**)
1	1	1	0	0	0	0
1	0	0	1	0	1	1
0	1	0	1	1	0	1
0	0	0	1	1	1	1

law as an example. Table 1.4 shows the truth values for all possible true–false combinations of **a** and **b**. It is clear that t(~(**a** & **b**)) and t(~**a** or ~**b**) are all equal.

1.2 PLAUSIBILITY LOGIC

Let us define the degree of plausibility of **b** given **a** as the function $\pi(\mathbf{b}|\mathbf{a})$. The proposition **a** is the premise that represents the context, and the proposition **b** is the target proposition of interest. The output of this function is a real number. By convention, let the highest plausibility of **b** given **a** be unity (this happens when **a** implies **b**) and let the lowest plausibility of **b** given **a** be zero (this happens when **a** implies ~**b**). As a result, $0 \leq \pi(\mathbf{b}|\mathbf{a}) \leq 1$. Moreover, $\pi(\mathbf{b}|\mathbf{a}) > \pi(\mathbf{c}|\mathbf{a})$ if **b** is more plausible than **c** given **a**. It is clear that logically equivalent propositions are equally plausible. For instance,

$$\pi(\sim\sim \mathbf{b} \mid \mathbf{a}) = \pi(\mathbf{b} \mid \mathbf{a}) \tag{1.7}$$

$$\pi(\mathbf{b} \,\&\, \mathbf{c} \mid \mathbf{a}) = \pi(\mathbf{c} \,\&\, \mathbf{b} \mid \mathbf{a}) \tag{1.8}$$

$$\pi(\mathbf{b} \,\&\, (\mathbf{c} \,\&\, \mathbf{d}) \mid \mathbf{a}) = \pi((\mathbf{b} \,\&\, \mathbf{c}) \,\&\, \mathbf{d} \mid \mathbf{a}) \tag{1.9}$$

1.2.1 Negation function and theorem

We postulate that $\exists N:[0,1]\rightarrow[0,1]$ such that $\pi(\sim\mathbf{b}|\mathbf{a}) = N[\pi(\mathbf{b}|\mathbf{a})]$, where N(.) is called the negative function (Cox 1946, 1961; Beck 2010). Note that the true–false status of ~**b**|**a** is completely known if the true-false status of **b**|**a** is known. This suggests that the plausibility of ~**b**|**a** should be completely known if the plausibility of **b**|**a** is known. That is, $\pi(\sim\mathbf{b}|\mathbf{a})$ is a function of $\pi(\mathbf{b}|\mathbf{a})$, and we denote this function as the negation function N(.). The negation function should have the following properties:

N1: N is continuous and decreasing.
N2: $N(1) = 0$, $N(0) = 1$
N3: $x = N[N(x)]\ \forall x \in [0,1]$
N1 and N2 are obvious. N3 is because $\pi(\mathbf{b}|\mathbf{a}) = \pi(\sim\sim\mathbf{b}|\mathbf{a}) = N[\pi(\sim\mathbf{b}|\mathbf{a})] = N\{N[\pi(\mathbf{b}|\mathbf{a})]\}$.

1.2.1.1 Negation theorem

N:[0,1]→[0,1] has properties N1~N3 if and only if $\exists\phi$:[0,1]→[0,1] with $\phi(x)$ continuous and increasing and $\phi(0) = 0$ & $\phi(1) = 1$ such that

$$N(x) = \phi^{-1}\left[1-\phi(x)\right] \quad \forall x \in [0,1] \tag{1.10}$$

Note the theorem states that the two propositions ("N:[0,1]→[0,1] has properties N1~N3" and "$\exists\phi$:[0,1]→[0,1] ... such that $N(x) = \phi^{-1}[1-\phi(x)]$ $\forall x\in[0,1]$") imply each other.

1.2.1.2 Proof of the negation theorem

The proof of the negation theorem can be found in Cox (1946, 1961). The proof of "the second proposition implying the first one" is simple. If $\phi(x)$ is continuous, increasing, and $\phi(0) = 0$ & $\phi(1) = 1$, it is clear that $N(x) = \phi^{-1}[1-\phi(x)]$ is continuous, decreasing, and $N(0) = 1$ & $N(1) = 0$. This shows that N1 and N2 hold. Moreover, $N[N(x)] = \phi^{-1}(1-\phi\{\phi^{-1}[1-\phi(x)]\}) = \phi^{-1}(1-[1-\phi(x)]) = \phi^{-1}[\phi(x)] = x$. This shows that N3 holds. The proof of "the first proposition implying the second one" is tricky. The key is to how that there exists a suitable $\phi(x)$ function such that N(x) can be always expressed as $\phi^{-1}[1-\phi(x)]$. Let us try the following candidate:

$$\phi(x) = \frac{1}{2}\left[1+x-N(x)\right] \tag{1.11}$$

Note that this particular $\phi(x)$ function is legitimate: it is continuous (because N is continuous), increasing (because N is decreasing), and also satisfies $\phi(0) = 0$ & $\phi(1) = 1$. Moreover,

$$\phi\left[N(x)\right] = \frac{1}{2}\left\{1+N(x)-N\left[N(x)\right]\right\} = \frac{1}{2}\left[1-x+N(x)\right] \tag{1.12}$$

It is then clear that

$$\phi(x)+\phi\left[N(x)\right] = \frac{1}{2}\left[1+x-N(x)\right]+\frac{1}{2}\left[1-x+N(x)\right] = 1 \tag{1.13}$$

As a result, $\phi[N(x)] = 1-\phi(x)$, hence, $N(x) = \phi^{-1}[1-\phi(x)]$. This shows that there indeed exists a $\phi(x)$ function (i.e., Equation 1.11) such that N(x) can be always expressed as $\phi^{-1}[1-\phi(x)]$. This is the end of proof.

Note that by replacing $x = \pi(\mathbf{b}|\mathbf{a})$ in Equation (1.13), we have

$$\phi\left[\pi(\mathbf{b}\mid\mathbf{a})\right]+\phi\left\{N\left[\pi(\mathbf{b}\mid\mathbf{a})\right]\right\} = \phi\left[\pi(\mathbf{b}\mid\mathbf{a})\right]+\phi\left[\pi(\sim\mathbf{b}\mid\mathbf{a})\right] = 1 \tag{1.14}$$

1.2.2 Conjunction function and theorem

We further postulate that $\exists C:[0,1]\times[0,1]\rightarrow[0,1]$ such that $\pi(\mathbf{b\&c|a}) = C[\pi(\mathbf{b|c\&a}),\pi(\mathbf{c|a})]$, where $C(.,.)$ is called the conjunction function (Cox 1946, 1961). Note that the true–false status of **b&c|a** is completely known if the true–false statuses of **c|a** and **b|c&a** are known. This suggests that the plausibility of **b&c|a** should be completely known if the plausibilities of **c|a** and **b|c&a** are known. That is, $\pi(\mathbf{b\&c|a})$ is a function of $\pi(\mathbf{b|c\&a})$ and $\pi(\mathbf{c|a})$, and we denote this function as the conjunction function $C(. , .)$ The conjunction function should have the following properties:

C1: $C(x,y) = C(y,x)\ \forall x,y\in[0,1]$
C2: C is continuous and increasing
C3: $x = C(1,x)\ \forall x\in[0,1]$
C4: $C[C(x,y),z] = C[x,C(y,z)]\ \forall x,y,z\in[0,1]$

C1 is intuitive and C2 is obvious. C3 holds because if $\mathbf{b}\equiv\mathbf{c}$, we have $\pi(\mathbf{b|a}) = C[\pi(\mathbf{b|b\&a}),\pi(\mathbf{b|a})] = C[1,\pi(\mathbf{b|a})]\times\pi(\mathbf{b|a})$. C4 holds because $\pi[\mathbf{b\&(c\&d)|a}] = C[\pi(\mathbf{b|c\&d\&a}),\pi(\mathbf{c\&d|a})] = C\{\pi(\mathbf{b|c\&d\&a}),C[\pi(\mathbf{c|d\&a}),\pi(\mathbf{d|a})]\}$ and $\pi[\mathbf{(b\&c)\&d|a}] = C[\pi(\mathbf{b\&c|d\&a}),\pi(\mathbf{d|a})] = C\{C[\pi(\mathbf{b|c\&d\&a}),\pi(\mathbf{c|d\&a})],\pi(\mathbf{d|a})\}$. The fact that $\pi[\mathbf{b\&(c\&d)|a}] = \pi[\mathbf{(b\&c)\&d|a}]$ suggests C4.

1.2.2.1 Conjunction theorem (stated without proof)

$C:[0,1]\times[0,1]\rightarrow[0,1]$ has properties C1~C4 if and only if $\exists\varphi:[0,1]\rightarrow[0,1]$ with $\varphi(x)$ continuous and increasing and $\varphi(0) = 0$ & $\varphi(1) = 1$ such that

$$C(x,y)=\varphi^{-1}\left[\varphi(x)\times\varphi(y)\right]\qquad\forall x,y\in[0,1]\tag{1.15}$$

Note that by replacing $x = \pi(\mathbf{b|c\&a})$ and $y = \pi(\mathbf{c|a})$ in Equation (1.15), we have

$$\varphi\{C[\pi(\mathbf{b}\mid\mathbf{c}\,\&\,\mathbf{a}),\pi(\mathbf{c}\mid\mathbf{a})]\}=\varphi[\pi(\mathbf{b}\mid\mathbf{c}\,\&\,\mathbf{a})]\times\varphi[\pi(\mathbf{c}\mid\mathbf{a})]\tag{1.16}$$

The proof of the conjunction theorem can be found in Cox (1946, 1961).

1.3 PROBABILITY LOGIC

Note that the φ function in the conjunction theorem can be different from the ϕ function in the negation theory although they share the same properties. Nonetheless, it is legitimate to choose the same function, that is, $\varphi = \phi$. In this case, we have

$$\phi[\pi(\mathbf{b}\,\&\,\mathbf{c}\mid\mathbf{a})]=\phi[\pi(\mathbf{b}\mid\mathbf{c}\,\&\,\mathbf{a})]\times\phi[\pi(\mathbf{c}\mid\mathbf{a})]\tag{1.17}$$

$$\phi[\pi(\sim \mathbf{b} \mid \mathbf{a})] + \phi[\pi(\mathbf{b} \mid \mathbf{a})] = 1 \tag{1.18}$$

If we further define "probability" as the following composite function: P(.) = $\phi[\pi(.)]$, we have

$$P(\mathbf{b} \,\&\, \mathbf{c} \mid \mathbf{a}) = P(\mathbf{b} \mid \mathbf{c} \,\&\, \mathbf{a}) \times P(\mathbf{c} \mid \mathbf{a}) \tag{1.19}$$

$$P(\sim \mathbf{b} \mid \mathbf{a}) + P(\mathbf{b} \mid \mathbf{a}) = 1 \tag{1.20}$$

where P(**b**|**a**) is the probability of **b** given **a**. Note that this probability is related to degree of belief (Bayesian), not related to relative frequency (frequentist).

1.4 AXIOMS FOR BAYESIAN PROBABILITY

Now, we are ready to write down the four Axioms of (Bayesian) probability:

P1:

$$P(\mathbf{b} \,\&\, \mathbf{c} \mid \mathbf{a}) = P(\mathbf{b} \mid \mathbf{c} \,\&\, \mathbf{a}) \cdot P(\mathbf{c} \mid \mathbf{a}) \tag{1.21}$$

P2:

$$P(\sim \mathbf{b} \mid \mathbf{a}) + P(\mathbf{b} \mid \mathbf{a}) = 1 \tag{1.22}$$

P3:

$$P(\mathbf{b} \mid \mathbf{a}) \geq 0 \tag{1.23}$$

P4:

$$P(\mathbf{b} \mid \mathbf{a} \,\&\, \mathbf{b}) = 1 \tag{1.24}$$

P1 and P2 are the same as Equations (1.19) and (1.20). P3 holds because P(**b**|**a**) = $\phi[\pi(\mathbf{b}|\mathbf{a})]$ is negative (ϕ has output space of [0,1]). P4 holds because P(**b**|**a**&**b**) = $\phi[\pi(\mathbf{b}|\mathbf{a}\&\mathbf{b})]$ = $\phi(1)$ = 1.

There are more properties that can be derived from Axioms P1 to P4:

P5:

$$\text{(a)}\ P(\sim \mathbf{b} \mid \mathbf{a} \,\&\, \mathbf{b}) = 0 \tag{1.25}$$

$$\text{(b)}\ P(\mathbf{b} \mid \mathbf{a}) \in [0,1] \tag{1.26}$$

Note that P5(a) follows P2 and P4. P5(b) follows P3 and P4.

P6:

(a) $P(\mathbf{c} \text{ or } \mathbf{b} \mid \mathbf{a}) = P(\mathbf{c} \mid \mathbf{a}) + P(\mathbf{b} \mid \mathbf{a}) - P(\mathbf{c} \,\&\, \mathbf{b} \mid \mathbf{a})$ (1.27)

(b) If **a** implies that **c&b** is false, that is, **b** and **c** are mutually exclusive:

$$P(\mathbf{c} \text{ or } \mathbf{b} \mid \mathbf{a}) = P(\mathbf{c} \mid \mathbf{a}) + P(\mathbf{b} \mid \mathbf{a}) \quad (1.28)$$

(c) if proposition **a** implies that only one of the propositions $\mathbf{b}_1, \mathbf{b}_2, \ldots, \mathbf{b}_N$ can be true, that is, they are mutually exclusive:

$$P(\mathbf{b}_1 \text{ or } \mathbf{b}_2 \text{ or } \ldots \text{ or } \mathbf{b}_N \mid \mathbf{a}) = \sum_{n=1}^{N} P(\mathbf{b}_n \mid \mathbf{a}) \quad (1.29)$$

If, in addition, proposition **a** implies that one must be true

$$\sum_{n=1}^{N} P(\mathbf{b}_n \mid \mathbf{a}) = 1 \quad (1.30)$$

Proof for P6(a)

From the second De Morgan's law, we have **c** or **b** ≡ ~(~**c** & ~**b**). Therefore,

$$P(\mathbf{c} \text{ or } \mathbf{b} \mid \mathbf{a}) = P(\sim(\sim \mathbf{c} \,\&\, \sim \mathbf{b}) \mid \mathbf{a}) \quad (1.31)$$

$= 1 - P(\sim\mathbf{c} \,\&\, \sim\mathbf{b}|\mathbf{a})$ (from P2)
$= 1 - P(\sim\mathbf{c}|\sim\mathbf{b} \,\&\, \mathbf{a})P(\sim\mathbf{b}|\mathbf{a})$ (from P1)
$= 1 - [1 - P(\mathbf{c}|\sim\mathbf{b} \,\&\, \mathbf{a})]P(\sim\mathbf{b}|\mathbf{a})$ (from P2)
$= 1 - P(\sim\mathbf{b}|\mathbf{a}) + P(\mathbf{c} \,\&\, \sim\mathbf{b}|\mathbf{a})$ (from P1)
$= P(\mathbf{b}|\mathbf{a}) + P(\sim\mathbf{b}|\mathbf{c} \,\&\, \mathbf{a})P(\mathbf{c}|\mathbf{a})$ (from P1, P2)
$= P(\mathbf{b}|\mathbf{a}) + [1 - P(\mathbf{b}|\mathbf{c} \,\&\, \mathbf{a})]P(\mathbf{c}|\mathbf{a})$ (from P2)
$= P(\mathbf{b}|\mathbf{a}) + P(\mathbf{c}|\mathbf{a}) - P(\mathbf{b} \,\&\, \mathbf{c}|\mathbf{a})$ (from P1)

P6(b) follows immediately because $P(\mathbf{c}\&\mathbf{b}|\mathbf{a}) = 0$. P6(c) can be proved by mathematical induction.

P7:

If proposition **a** implies that one, and only one, of the propositions $\mathbf{b}_1, \mathbf{b}_2, \ldots, \mathbf{b}_N$ is true,

(a) $P(\mathbf{c} \mid \mathbf{a}) = \sum_{n=1}^{N} P(\mathbf{c} \,\&\, \mathbf{b}_n \mid \mathbf{a})$ [Marginalization Theorem] (1.32)

(b) $P(\mathbf{c} \mid \mathbf{a}) = \sum_{n=1}^{N} P(\mathbf{c} \mid \mathbf{b}_n \,\&\, \mathbf{a}) P(\mathbf{b}_n \mid \mathbf{a})$ [Total Probability Theorem]

(1.33)

$$(c)\ P(\mathbf{b}_k \mid \mathbf{c} \,\&\, \mathbf{a}) = \frac{P(\mathbf{c} \mid \mathbf{b}_k \,\&\, \mathbf{a})P(\mathbf{b}_k \mid \mathbf{a})}{\sum_{n=1}^{N} P(\mathbf{c} \mid \mathbf{b}_n \,\&\, \mathbf{a})P(\mathbf{b}_n \mid \mathbf{a})} \quad [\text{Bayes Theorem}] \tag{1.34}$$

Proof for P7

Let $\mathbf{b} \equiv \mathbf{b}_1$ or $\mathbf{b}_2$ or … or $\mathbf{b}_N$. Because **a** implies **b** is true, $\mathbf{a} \equiv \mathbf{a}\&\mathbf{b}$, so from P4, $P(\mathbf{b}|\mathbf{a}) = P(\mathbf{b}|\mathbf{a}\&\mathbf{b}) = 1$ and $P(\mathbf{c}\&\mathbf{b}|\mathbf{a}) = P(\mathbf{c}|\mathbf{b}\&\mathbf{a}) \times P(\mathbf{b}|\mathbf{a}) = P(\mathbf{c}|\mathbf{a})$, so

$$\begin{aligned} P(\mathbf{c} \mid \mathbf{a}) &= P(\mathbf{c} \,\&\, (\mathbf{b}_1 \text{ or } \dots \text{ or } \mathbf{b}_N) \mid \mathbf{a}) \\ &= P((\mathbf{c} \,\&\, \mathbf{b}_1) \text{ or } \dots \text{ or } (\mathbf{c} \,\&\, \mathbf{b}_N) \mid \mathbf{a}) \end{aligned} \tag{1.35}$$

Since **a** implies that $\mathbf{b}_m$ and $\mathbf{b}_n$ cannot both be true if $m \neq n$, it must be that $(\mathbf{c}\&\mathbf{b}_m)$ & $(\mathbf{c}\&\mathbf{b}_n) \equiv \mathbf{c}$ & $(\mathbf{b}_m\&\mathbf{b}_n)$ is false, that is, $(\mathbf{c}\&\mathbf{b}_m)$ and $(\mathbf{c}\&\mathbf{b}_n)$ are mutually exclusive. Due to P6(c), Equation (1.32) holds. P7(b) follows from P7(a) using P1. From P7(b), the denominator on the right-hand side of Equation (1.34) is equal to $P(\mathbf{c}|\mathbf{a})$, so we only need to show that $P(\mathbf{b}_k|\mathbf{c}\&\mathbf{a}) \times P(\mathbf{c}|\mathbf{a}) = P(\mathbf{c}|\mathbf{b}_k\&\mathbf{a}) \times P(\mathbf{b}_k|\mathbf{a})$. This follows from P1 because both are equal to $P(\mathbf{b}_k\&\mathbf{c}|\mathbf{a}) = P(\mathbf{c}\&\mathbf{b}_k|\mathbf{a})$.

1.5 CONTINUOUS VARIABLE

Thus far, the focus has been on propositions. However, continuous variables are usually of concern in geotechnical site characterization. Let us denote a capital letter (such as X) as a random variable and denote a small letter (such as x) as a fixed value. Now consider the following propositions related to continuous random variables:

a = "certain context holds"
b = "$x \leq X < x + dx$"
c = "$y \leq Y < y + dy$"

where dx and dy are infinitesimal increments of x and y. In this case, P1 becomes

$$\begin{aligned} &P(\text{"}x \leq X < x + dx\text{"} \,\&\, \text{"}y \leq Y < y + dy\text{"} \mid \mathbf{a}) \\ &= P(\text{"}x \leq X < x + dx\text{"} \mid \text{"}y \leq Y < y + dy\text{"} \,\&\, \mathbf{a}) \times P(\text{"}y \leq Y < y + dy\text{"} \mid \mathbf{a}) \end{aligned} \tag{1.36}$$

Note that all probabilities in Equation (1.36) are infinitesimally small, and P1 becomes not useful.

To make P1 useful, let us define the probability density function (PDF) of Y by f(y|a):

$$f(y \mid a) = \lim_{dy \to 0} \frac{P(\text{"}y \le Y \sim\ < y + dy\text{"} \mid a)}{dy} \tag{1.37}$$

Similarly, the conditional PDF of X given Y can be defined as

$$f(x \mid y, a) = \lim_{dx \to 0} \frac{P(\text{"}x \le X < x + dx\text{"} \mid \text{"}y \le Y < y + dy\text{"}\ \&\ a)}{dx} \tag{1.38}$$

We can also further define the bivariate (joint) PDF of X and Y:

$$f(x, y \mid a) = \lim_{dx \to 0, dy \to 0} \frac{P(\text{"}x \le X < x + dx\text{"}\ \&\ \text{"}y \le Y < y + dy\text{"} \mid a)}{dx \times dy} \tag{1.39}$$

With the definition of the PDFs, the original P1 becomes

$$f(x, y \mid a)dxdy = f(x \mid y, a)dx \times f(y \mid a)dy \tag{1.40}$$

or simply

P1

$$f(x, y \mid a) = f(x \mid y, a) \times f(y \mid a) \tag{1.41}$$

The original P3 (Equation 1.23) suggests that the PDF must be non-negative:

P3

$$f(x \mid a) \ge 0 \tag{1.42}$$

However, the PDF does not need to be less than 1 (probability has an upper bound of 1, but probability density does not).

For P7, let us consider propositions $\mathbf{b}_1, \mathbf{b}_2, \ldots, \mathbf{b}_N$ ($N \to \infty$), where $\mathbf{b}_n$ = "$x_n \le X < x_n + dx$", $x_n + dx = x_{n+1}$, $x_1 = -\infty$, and $x_{N+1} = \infty$. It is clear that $\mathbf{b}_1, \mathbf{b}_2, \ldots, \mathbf{b}_N$ satisfy that one, and only one, of them must be true. Let proposition $\mathbf{c}$ = "$y \le Y < y + dy$". The original P7(a) (Equation 1.32) becomes

$$f(y \mid a)dy = \lim_{N \to \infty} \sum_{n=1}^{N} f(x_n, y \mid a)dxdy = \int_{-\infty}^{\infty} f(x, y \mid a)dxdy \tag{1.43}$$

or simply

P7(a)

$$f(y \mid a) = \int_{-\infty}^{\infty} f(x, y \mid a)dx \text{ [Marginalization Theorem]} \tag{1.44}$$

The total probability theorem and Bayes theorem for continuous variables then follow:

P7(b)

$$f(y \mid a) = \int f(y \mid x, a)f(x \mid a)dx \text{ [Total Probability Theorem]} \tag{1.45}$$

P7(c)

$$f(x \mid y, a) = \frac{f(y \mid x, a)f(x \mid a)}{\int f(y \mid x, a)f(x \mid a)dx} \text{ [Bayes Theorem]} \tag{1.46}$$

Chapter 2

Review of probability and models

In this chapter, some definitions and concepts related to continuous random variables are reviewed in the context of the Bayesian notion of probability, including probability density function (PDF), cumulative distribution function (CDF), quantile function, expectation, correlation, and independence. Next, some useful PDF models for Bayesian analysis are introduced. In the previous chapter, there is always a contextual proposition **a** = "certain context holds" for probability. In this chapter, this contextual proposition will be omitted, that is, f(x|**a**) will be denoted by f(x), when there is no confusion. There is always a certain context although we do not show **a**.

2.1 DEFINITIONS AND CONCEPTS RELATED TO CONTINUOUS RANDOM VARIABLES

2.1.1 Probability density function

A function f(x) is a PDF of X if and only if it satisfies the following two properties:

$$1.\ f(x) \geq 0 \quad \forall x \tag{2.1}$$

$$2.\ \int_{-\infty}^{\infty} f(x)\,dx = 1 \tag{2.2}$$

The second condition is due to the fact that

$$3.\ P(a \leq X \leq b) = \int_{a}^{b} f(x)\,dx \tag{2.3}$$

Therefore,

$$4.\ P(-\infty \leq X \leq \infty) = \int_{-\infty}^{\infty} f(x)\,dx = 1 \tag{2.4}$$

DOI: 10.1201/9781003309765-2

2.1.2 Cumulative distribution function

The CDF, denoted by F(x), is defined as $P(X \le x)$. Note that F(x) is a probability, so $0 \le F(x) \le 1\ \forall x$. According to Equation (2.3),

$$F(x) = P(X \le x) = \int_{-\infty}^{x} f(\tau)\,d\tau \tag{2.5}$$

where τ is a dummy variable. It is clear that CDF F(x) is the integration function of PDF f(x). Conversely, f(x) is the differentiation function of F(x), that is, f(x) = d[F(x)]/dx. Figure 2.1 shows the relationship between F(x) and f(x). The height of F(x) at x_0 is equal to the area under f(x) from $-\infty$ to x_0 (shaded area).

A CDF must satisfy the following properties:

1. $F(-\infty) = 0$ (2.6)
2. $F(\infty) = 1$ (2.7)
3. $F(x)$ is non – decreasing (2.8)

The first two properties are due to the fact that $P(X \le -\infty) = 0$ and $P(X \le \infty) = 1$. The third property is because PDF is the differentiation function of CDF and PDF is non-negative.

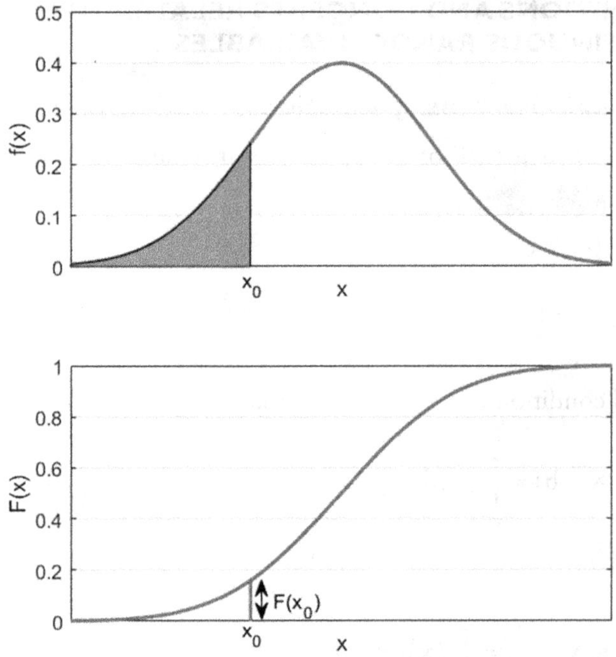

Figure 2.1 Relationship between F(x) and f(x).

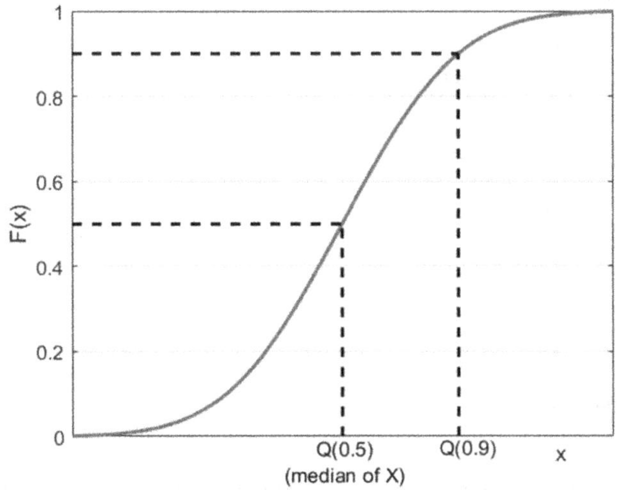

Figure 2.2 Significance of the quantile function.

2.1.3 Quantile function and Monte Carlo simulation

The quantile function, denoted by Q(p) ($0 \le p \le 1$), outputs the x value such that $P(X \le x) = p$. It is clear that the Q function is identical to the inverse of the CDF, that is, $Q(.) = F^{-1}(.)$. The number Q(p) is called the p-fractile. The 0.5-fractile of X = Q(0.5) is called the median value of X because P(X ≤ median value) = $P[X \le Q(0.5)] = 0.5$. The 0.9-fractile of X = Q(0.9) is on the high side of X, as shown in Figure 2.2. There is a chance of 0.9 that X is less or equal to the 0.9-fractile.

If the quantile function of X is known, the Monte Carlo simulation (MCS) can be conducted by the following procedure to draw random samples of X: (a) Let U be a random sample uniformly distributed over the interval [0, 1]; (b) a sample of X can be obtained as X = Q(U).

2.1.3.1 Proof for the procedure of MCS

Let Y = Q(U). The key is to show that Y follows the same probability distribution as X. To see this, consider the following equations:

$$\begin{aligned} P(Y \le x) &= P\left[Q(U) \le x\right] = P\left[F^{-1}(U) \le x\right] = P\left\{F\left[F^{-1}(U)\right] \le F(x)\right\} \\ &= P\left[U \le F(x)\right] \end{aligned} \tag{2.9}$$

Because U is uniformly distributed over [0,1], $P[U \le F(x)] = F(x)$. Therefore, we have

$$P(Y \le x) = F(x) \tag{2.10}$$

Namely, Y follows the same probability distribution as X. This is the end of the proof. An important implication of this proof is that F(X) = U is always uniformly distributed over [0,1].

2.1.4 Expected value

From the frequentist perspective, the expected value is the arithmetic mean of a large number of independently selected outcomes of a random variable. The expected value of a function of X, denoted by E[g(X)], can be computed by the following integral:

$$E\left[g(X)\right] = \int_{-\infty}^{\infty} g(x) \cdot f(x)dx \tag{2.11}$$

From the Bayesian perspective, the expected value is the average value of a random variable weighted by its probability density, which can be computed using the same formula (Equation 2.11). There are several important special cases of expected values depending on the g(.) function.

2.1.4.1 *Mean value of X*

The mean (average) of X, denoted by μ = E(X), can be computed as

$$\mu = E(X) = \int_{-\infty}^{\infty} x \cdot f(x)dx \tag{2.12}$$

The mean value of X is the centroid location of the PDF of X. In general, the mean value is not the same as the median value (0.5-fractile). They are the same if and only if the PDF is symmetric about its mean value.

2.1.4.2 *Variance of X*

The variance of X, denoted by σ^2 = Var(X), can be computed as

$$\sigma^2 = Var(X) = E\left[(X-\mu)^2\right] = \int_{-\infty}^{\infty} (x-\mu)^2 \cdot f(x)dx \tag{2.13}$$

The variance of X quantifies the uncertainty in the PDF of X (uncertainty is significant if the variance is large). The following identity is popular for the calculation of Var(X):

$$Var(X) = E\left(X^2\right) - \mu^2 \tag{2.14}$$

where $E(X^2)$ is called the second moment of X:

$$E(X^2) = \int_{-\infty}^{\infty} x^2 \cdot f(x)dx \tag{2.15}$$

2.1.4.3 *Proof of Equation (2.14)*

$$\begin{aligned} Var(X) &= E\left[(X-\mu)^2\right] = E\left(X^2 - 2\mu X + \mu^2\right) = E\left(X^2\right) - 2\mu E(X) \\ &\quad + \mu^2 = E\left(X^2\right) - \mu^2 \end{aligned} \tag{2.16}$$

Note that the unit of Var(X) has unclear physical meaning, for example, if X has the unit of meter, Var(X) has the unit of m^2. To resolve this issue, the standard deviation, denoted by σ, is defined as the square root of Var(X). The standard deviation quantifies the width of the PDF of X. The coefficient of variation (COV) of X is defined as the ratio between the standard deviation and the mean:

$$COV = \frac{\sigma}{\mu} \tag{2.17}$$

The COV can be considered as the normalized standard deviation: a COV of 30% means that the standard deviation is equal to 30% of the mean value.

2.1.4.4 *Covariance between X and Y*

The covariance between two random variables X and Y, denoted by COV(X,Y), can be computed as

$$\begin{aligned} COV(X,Y) &= E\left[(X-\mu_X)(Y-\mu_Y)\right] \\ &= \int_{-\infty}^{\infty}\int_{-\infty}^{\infty} (x-\mu_X)(y-\mu_Y) \cdot f(x,y)dxdy \end{aligned} \tag{2.18}$$

where μ_X is the mean value of X, μ_Y is the mean value of Y, and f(x,y) is the bivariate (joint) PDF of X and Y. The following identity is popular for the calculation of Var(X):

$$COV(X,Y) = E(XY) - \mu_X\mu_Y \tag{2.19}$$

2.1.4.5 *Proof of Equation (2.19)*

$$\begin{aligned} COV(X,Y) &= E\left[(X-\mu_X)(Y-\mu_Y)\right] = E\left(XY - \mu_X Y - \mu_Y X + \mu_X\mu_Y\right) \\ &= E(XY) - \mu_X E(Y) - \mu_Y E(X) + \mu_X\mu_Y = E(XY) - \mu_X\mu_Y \end{aligned} \tag{2.20}$$

It is clear that the covariance between X and X itself is the variance:

$$COV(X,X) = Var(X) \tag{2.21}$$

The covariance between X and Y quantifies the degree-of-linear correlation between X and Y. If COV(X,Y) is positive (positive correlation), X and Y have positive (linear) correlation, for example, Figure 2.3(a). If COV(X,Y) is negative (negative correlation), X and Y have a negative (linear) correlation, for example, Figure 2.3(b). If COV(X,Y) = 0, X and Y have no (linear) correlation, for example, Figure 2.3(c).

Note that the unit of COV(X,Y) has unclear physical meaning, for example, if X and Y have the units of kilogram and meter, COV(X,Y) has the unit of kg×m. To resolve this issue, the Pearson product moment correlation coefficient (or simply correlation coefficient) between X and Y, denoted by ρ(X,Y), can be computed as

$$\rho(X,Y) = \frac{COV(X,Y)}{\sigma_X \cdot \sigma_Y} \tag{2.22}$$

where σ_X is the standard deviation of X, and σ_Y is the standard deviation of Y. The correlation coefficient can be considered as the normalized covariance. It can be shown that $-1 \leq \rho(X,Y) \leq 1$, where $\rho(X,Y) = 1$ corresponds to perfect positive correlation, and $\rho(X,Y) = -1$ corresponds to perfect negative correlation.

2.1.4.6 Identities for expected values

There are some identities for expected values. First, the expectation operation is linear:

$$E(aX + bY + c) = a \cdot E(X) + b \cdot E(Y) + c \tag{2.23}$$

For the variance, the rule is slightly different:

$$Var(aX + bY + c) = a^2 \cdot Var(X) + b^2 \cdot Var(Y) + 2ab \cdot COV(X,Y) \tag{2.24}$$

The constant shift term "c" has no effect on the variance because it does not affect the uncertainty. The covariance is bilinear, so

$$COV(a_1X + b_1Y + c_1, a_2X + b_2Y + c_2) = a_1a_2 \cdot Var(X) + b_1b_2 \cdot Var(Y) + (a_1b_2 + a_2b_1) \cdot COV(X,Y) \tag{2.25}$$

Again, the constant shift terms c_1 and c_2 have no effect on the covariance.

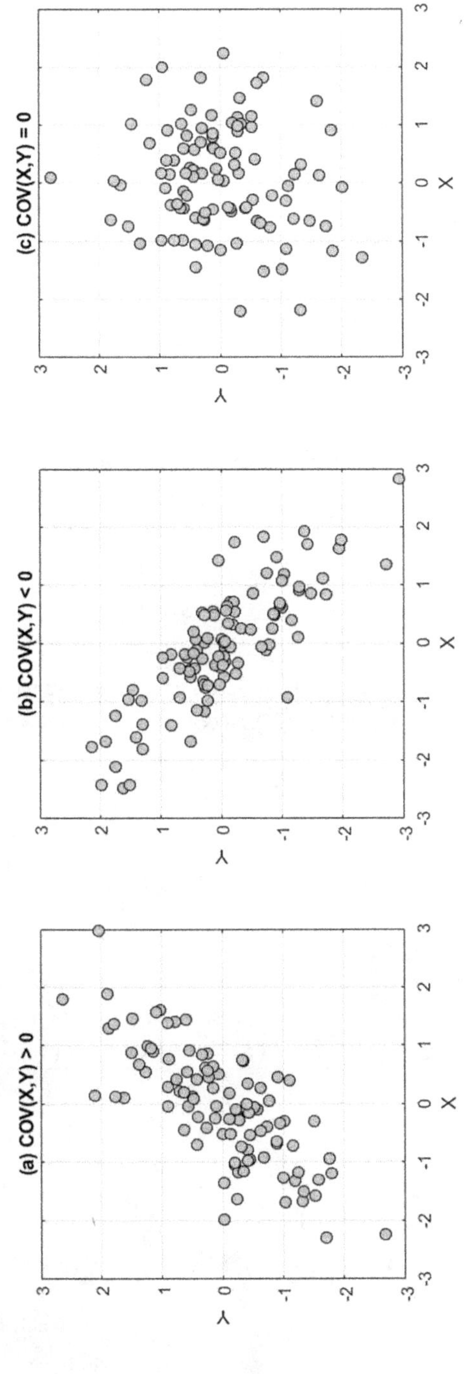

Figure 2.3 Significance of the sign of COV(X,Y).

2.1.4.7 Law of large numbers

The Law of Large Numbers (LLNs) states that the average of the results obtained from a large number of independent and identically distributed (i.i.d.) samples is close to the expected value and tends to become closer to the expected value as more samples are available. If $X^{(1)}$, $X^{(2)}$, ..., $X^{(m)}$ are i.i.d. samples of X, according to the LLN,

$$E[g(X)] = \int_{-\infty}^{\infty} g(x)\cdot f(x)dx \approx \frac{1}{m}\sum_{k=1}^{m} g\left(X^{(k)}\right) \tag{2.26}$$

Moreover, the approximate equality sign tends to become the equality sign as m approaches infinity. The following example shows how to estimate the expected values related to X and Y based on the LLN.

EXAMPLE 2.1 Estimation of expected values based on LLN

Given the following pairwise data for X and Y:

$$\begin{matrix} X = \\ Y = \end{matrix} \begin{bmatrix} 1 & 2 & 4 & 3 & 0 \\ 2 & 3 & 5 & 5 & 0 \end{bmatrix} \tag{2.27}$$

The purpose is to estimate some expected values related to X and Y, including μ_X, μ_Y, σ_X, σ_Y, COV(X,Y), and $\rho(X,Y)$. According to LLN,

$$\mu_X \approx \frac{1+2+4+3+0}{5} = 2,\ \mu_Y \approx \frac{2+3+5+5+0}{5} = 3 \tag{2.28}$$

$$\begin{aligned} \mathrm{Var}(X) &= E\left[(X-\mu_X)^2\right] \approx \frac{(1-2)^2+(2-2)^2+(4-2)^2+(3-2)^2+(0-2)^2}{5} \\ &= \frac{1+0+4+1+4}{5} = 2 \end{aligned} \tag{2.29}$$

$$\begin{aligned} \mathrm{Var}(Y) &= E\left[(Y-\mu_Y)^2\right] \approx \frac{(2-3)^2+(3-3)^2+(5-3)^2+(5-3)^2+(0-3)^2}{5} \\ &= \frac{1+0+4+4+9}{5} = \frac{18}{5} \end{aligned} \tag{2.30}$$

$$\begin{aligned} \mathrm{COV}(X,Y) &= E\left[(X-\mu_X)(Y-\mu_Y)\right] \\ &\approx \frac{(1-2)(2-3)+(2-2)(3-3)+(4-2)(5-3)+(3-2)(5-3)+(0-2)(0-3)}{5} \\ &= \frac{1+0+4+2+6}{5} = \frac{13}{5} \end{aligned} \tag{2.31}$$

$$\rho(X,Y)=\frac{COV(X,Y)}{\sigma_X\cdot\sigma_Y}\approx\frac{13/5}{\sqrt{2}\cdot\sqrt{18/5}}=0.969 \tag{2.32}$$

2.1.5 Independence

Two random variables are independent if the realization of one does not affect the probability distribution of the other. Formally, two random variables X and Y are independent ($X\perp Y$) if and only if their joint PDF is equal to the product of their marginal PDFs:

$$f(x,y)=f(x)\times f(y) \tag{2.33}$$

where f(x,y) is the bivariate (joint) PDF of X and Y; f(x) and f(y) are the marginal PDFs of X and Y. It is clear that Equation (2.33) is equivalent to the following two equations:

$$f(x\mid y)=f(x) \tag{2.34}$$

$$f(y\mid x)=f(y) \tag{2.35}$$

The concept of independence ($X\perp Y$) is similar to the concept of no correlation [COV(X,Y) = 0]. However, it can be shown that $X\perp Y$ implies COV(X,Y) = 0, but the converse is not true.

2.1.5.1 Proof for X⊥Y → COV(X,Y) = 0

Suppose that $X\perp Y$, which means that $f(x,y) = f(x) \times f(y)$. Let us consider the following equations:

$$COV(X,Y)=E(XY)-\mu_X\mu_Y=\int_{-\infty}^{\infty}\int_{-\infty}^{\infty}xy\cdot f(x,y)dxdy-\mu_X\mu_Y$$

$$=\int_{-\infty}^{\infty}\int_{-\infty}^{\infty}xy\cdot f(x)f(y)dxdy-\mu_X\mu_Y=\left(\int_{-\infty}^{\infty}x\cdot f(x)dx\right)\left(\int_{-\infty}^{\infty}y\cdot f(y)dy\right)-\mu_X\mu_Y$$

$$=\mu_X\mu_Y-\mu_X\mu_Y=0 \tag{2.36}$$

Therefore, $X\perp Y$ implies COV(X,Y) = 0. The proof for COV(X,Y) = 0 not implying $X\perp Y$ is straightforward: it can be proved by simply finding a counterexample with COV(X,Y) = 0 but with dependent (X,Y).

2.2 SOME UNIVARIATE PDF MODELS

2.2.1 Uniform PDF model

The uniform distribution, denoted by unif(a,b), is probably the simplest PDF model. The uniform PDF can be expressed as

$$f(x)=\begin{cases}\dfrac{1}{b-a} & a\le x\le b\\ 0 & o/w(\text{otherwise})\end{cases} \tag{2.37}$$

where a and b are the lower and upper bounds of X. Figure 2.4 shows some uniform PDFs. The CDF for the uniform PDF is

$$F(x)=\begin{cases}0 & x<a\\ \dfrac{x-a}{b-a} & a\le x\le b\\ 1 & b<x\end{cases} \tag{2.38}$$

If X ~ unif(a,b) (the sign "~" means "distributed as"), E(X) = μ = (a+b)/2 and Var(X) = σ^2 = $(b-a)^2/12$.

A special case for the uniform PDF is a = 0 and b = 1, called the standard uniform PDF. The corresponding random variable is denoted by U ~ unif(0,1). The relationship between a non-standard uniform X and the standard uniform U is as follows:

$$X=a+(b-a)\cdot U \tag{2.39}$$

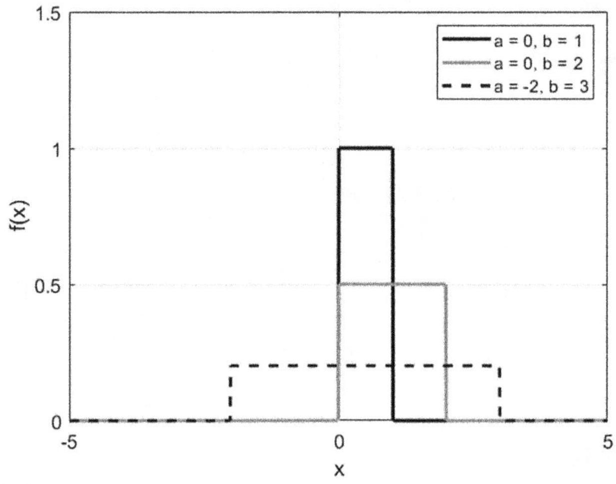

Figure 2.4 Some uniform PDFs.

Equation (2.39) also implies that the quantile function for the uniform PDF is simply Q(p) = a+(b–a)×p. Pseudo-random samples of U ~ unif(0,1) can be simulated by many programming platforms, such as Fortran, MATLAB, and Python. In MATLAB, the command "rand" will simulate a U sample. To simulate X ~ unif(a, b) in MATLAB, simply do the following two steps: (a) U = rand; (b) X = a + (b–a)×U.

2.2.2 Normal (Gaussian) PDF model

The normal (Gaussian) distribution, denoted by $N(\mu, \sigma^2)$, is one of the most popular PDF models. It is popular because many real data exhibit bell-shaped histograms, and this bell shape may be captured by the Gaussian function. The normal PDF can be expressed as

$$f(x) = \frac{1}{\sqrt{2\pi}}\frac{1}{\sigma}\exp\left[\frac{-(x-\mu)^2}{2\cdot\sigma^2}\right] \tag{2.40}$$

where $\mu = E(X)$ and $\sigma^2 = Var(X)$. Figure 2.5 shows some normal PDFs with various μ and σ^2. Note that the normal PDF is symmetric about its mean value. The CDF for the normal PDF is

$$F(x) = \int_{-\infty}^{x}\frac{1}{\sqrt{2\pi}\cdot\sigma}\exp\left[\frac{-(\tau-\mu)^2}{2\cdot\sigma^2}\right]\cdot d\tau \tag{2.41}$$

Note that the above integration does not have an analytical form.

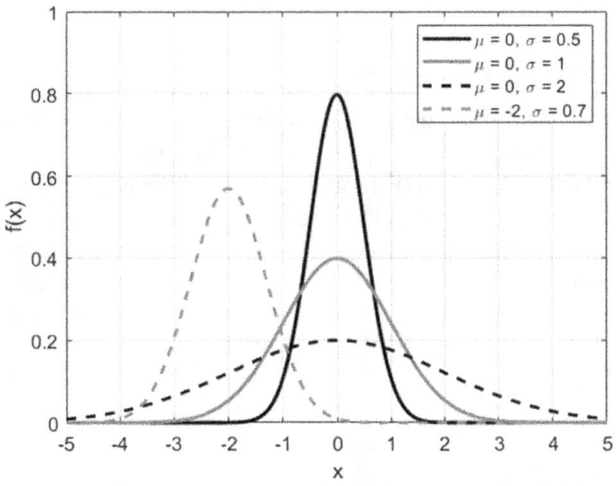

Figure 2.5 Some normal PDFs.

A special case for the normal PDF is $\mu = 0$ and $\sigma = 1$, called the standard normal PDF. The corresponding random variable is denoted by $Z \sim N(0,1)$. We adopt the $\varphi(z)$ function to denote its PDF:

$$\varphi(z) = \frac{1}{\sqrt{2\pi}} \exp\left(\frac{-z^2}{2}\right) \tag{2.42}$$

and we adopt the $\Phi(z)$ function to denote its CDF:

$$\Phi(z) = \int_{-\infty}^{z} \frac{1}{\sqrt{2\pi}} \exp\left(\frac{-\tau^2}{2}\right) \cdot d\tau \tag{2.43}$$

Again, the $\Phi(z)$ function does not have an analytical expression. Nonetheless, many programming platforms have built-in functions for the standard normal CDF, for example, $\Phi(z)$ can be evaluated by normcdf(z) in MATLAB. By definition, the quantile function for the standard normal is the inversion of its CDF: $Q(p) = \Phi^{-1}(p)$. Similarly, many programming platforms have built-in functions for the standard normal quantile function, for example, $\Phi^{-1}(p)$ can be evaluated by norminv(p) in MATLAB. To simulate $Z \sim N(0,1)$ in MATLAB, simply do the following two steps: (a) U = rand; (b) Z = norminv(U), or simply use the short-cut function Z = randn in one step. The relationship between a non-standard normal X and the standard normal Z is as follows:

$$X = \mu + \sigma \cdot Z \tag{2.44}$$

As a result, to simulate $X \sim N(\mu, \sigma^2)$ in MATLAB, simply do the following two steps: (a) Z = randn; (b) $X = \mu + \sigma \times Z$.

2.2.3 Lognormal PDF model

A potential issue for the normal PDF is that a normal random variable can be negative. However, many geotechnical parameters are non-negative, so the use of the normal PDF model can be controversial in this circumstance. A possible solution is to adopt the lognormal distribution, denoted by $LN(m, s^2)$. The lognormal random variable is derived from the normal random variable. If $Y \sim N(m, s^2)$ and $X = \exp(Y)$, X is lognormal, denoted by $X \sim LN(m, s^2)$. Note that the two parameters m and s^2 are the mean value and variance of ln(X), respectively. If μ and σ^2 are the mean and variance of X, the relationships between (μ, σ^2) and (m, s^2) are

$$m = \ln\left(\frac{\mu}{\sqrt{1+(\sigma/\mu)^2}}\right) \quad s = \sqrt{\ln\left[1+(\sigma/\mu)^2\right]} \tag{2.45}$$

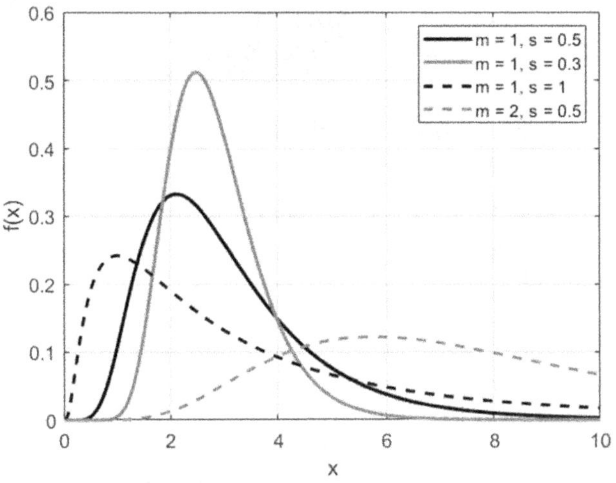

Figure 2.6 Some lognormal PDFs.

The lognormal PDF has the following expression:

$$f(x) = \frac{1}{\sqrt{2\pi}} \frac{1}{x} \frac{1}{s} \exp\left\{-\frac{1}{2s^2}\left[\ln(x) - m\right]^2\right\} \tag{2.46}$$

Figure 2.6 shows some lognormal PDFs with various m and s^2. Note that the lognormal PDF is no longer symmetric.

2.2.3.1 *Proof for the lognormal PDF in Equation (2.46)*

Let us start from the CDF of X:

$$F(x) = P(X \le x) = P\left[\exp(Y) \le x\right] = P\left[Y \le \ln(x)\right] = \int_{-\infty}^{\ln(x)} \frac{1}{\sqrt{2\pi}} \frac{1}{s} e^{-\frac{1}{2s^2}(y-m)^2} dy \tag{2.47}$$

Recall that the PDF is the differentiation function of the CDF. Therefore,

$$\begin{aligned} f(x) = \frac{dF(x)}{dx} &= \frac{d\left[\int_{-\infty}^{\ln(x)} \frac{1}{\sqrt{2\pi}} \frac{1}{s} e^{-\frac{1}{2s^2}(y-m)^2} dy\right]}{d\ln(x)} \cdot \frac{d\ln(x)}{dx} \\ &= \frac{1}{\sqrt{2\pi}} \frac{1}{x} \frac{1}{s} \exp\left\{-\frac{1}{2s^2}\left[\ln(x) - m\right]^2\right\} \end{aligned} \tag{2.48}$$

To simulate X ~ LN(m, s^2) in MATLAB, simply do the following three steps: (a) Z = randn; (b)Y = m + s×Z; (c) X = exp(Y).

2.2.4 Johnson system of distributions

Phoon (2008) and Phoon and Ching (2013) highlighted that the lognormal distribution is a member of a more general Johnson system (Johnson 1949), which can be expressed in the following form following the notations presented by Slifker and Shapiro (1980):

$$\frac{Z-\gamma}{\eta}=\kappa\left(\frac{X-\varepsilon}{\lambda}\right)=\kappa(X_n) \tag{2.49}$$

where (η, γ, λ, ε) are the four Johnson parameters; Z is standard normal; X_n = (X−ε)/λ is the normalized X. The SU member is unbounded ("U" denotes unbounded) and it is defined by

$$\kappa(X_n)=\sinh^{-1}(X_n)=\ln\left(X_n+\sqrt{1+X_n^2}\right) \tag{2.50}$$

The SB member is bounded between [ε, ε+λ] ("B" denotes bounded) and it is defined by

$$\kappa(X_n)=\ln\left(\frac{X_n}{1-X_n}\right) \tag{2.51}$$

The SL member (which is the shifted lognormal member) is bounded from *below* by ε ("L" denotes lognormal) and it is defined by

$$\kappa(X_n)=\ln(X_n) \tag{2.52}$$

Their PDFs are as follows:

$$f(x)=\begin{cases}\eta/\lambda\cdot\exp\left(-0.5\left[\gamma+\eta\sinh^{-1}(x_n)\right]^2\right)\Big/\sqrt{2\pi\left(1+x_n^2\right)} & \text{for SU}\\ \eta/\lambda\cdot\exp\left(-0.5\left(\gamma+\eta\ln\left[x_n/(1-x_n)\right]\right)^2\right)\Big/\left[\sqrt{2\pi}\cdot x_n(1-x_n)\right] & \text{for SB}\\ \eta\cdot\exp\left(-0.5\left[\gamma+\eta\ln(x-\varepsilon)\right]^2\right)\Big/\left[\sqrt{2\pi}\cdot(x-\varepsilon)\right] & \text{for SL}\end{cases} \tag{2.53}$$

Figure 2.7 shows some PDFs in the Johnson system – the Johnson system can generate PDFs with a wide range of shapes. For each of the SU, SB, and SL distributions, a baseline case is plotted. Then, the effect of each of the parameters (η, γ, λ, ε) is shown in the figure.

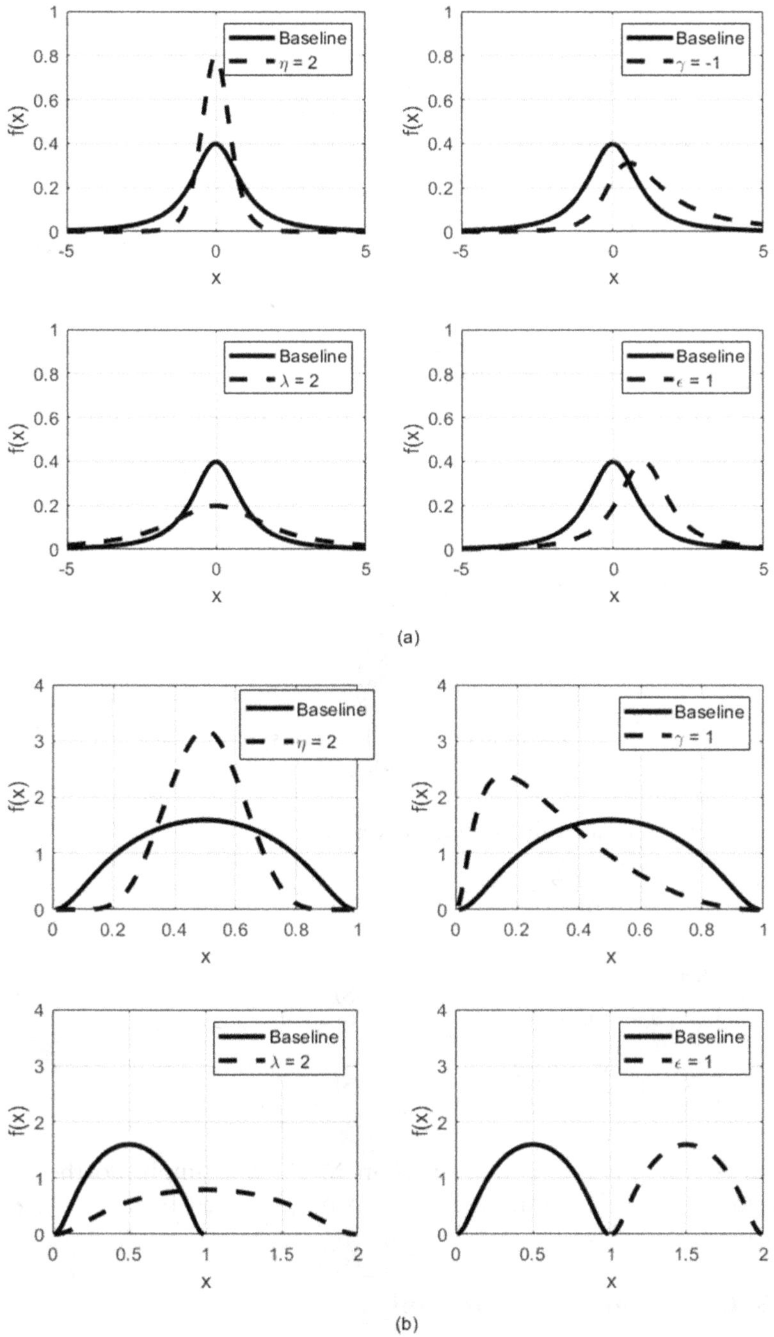

Figure 2.7 Some PDFs in the Johnson system: (a) SU distribution (baseline is with $\eta = 1, \gamma = 0; \lambda = 1; \varepsilon = 0$); (b) SB (baseline is with $\eta = 1, \gamma = 0; \lambda = 1; \varepsilon = 0$);

(Continued)

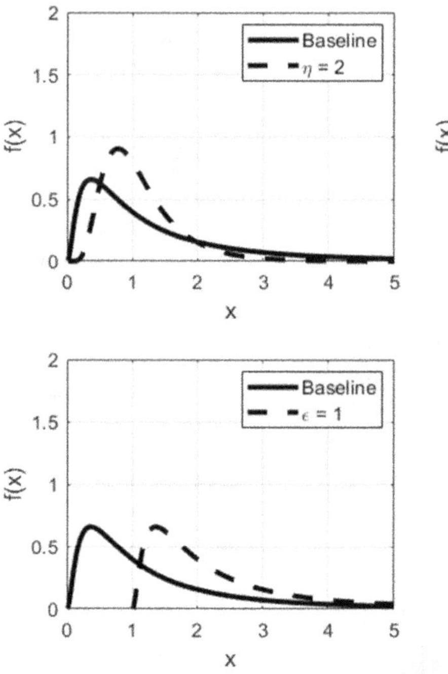

(c)

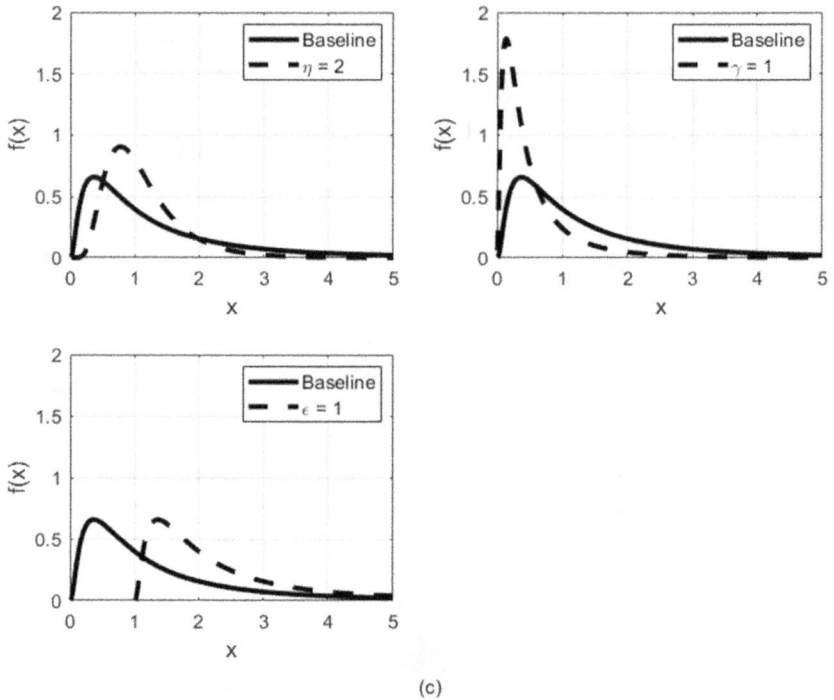

Figure 2.7 (Continued) Some PDFs in the Johnson system: (c) SL (baseline is with $\eta = 1, \gamma = 0; \varepsilon = 0$).

The inverse relationships between X and Z are

$$X = \begin{cases} \varepsilon + \lambda \times \sinh\left(\dfrac{Z-\gamma}{\eta}\right) & \text{SU} \\ \dfrac{\varepsilon + (\lambda + \varepsilon) \times \exp[(Z-\gamma)/\eta]}{1 + \exp[(Z-\gamma)/\eta]} & \text{SB} \\ \varepsilon + \exp\left(\dfrac{Z-\gamma}{\eta}\right) & \text{SL} \end{cases} \tag{2.54}$$

Therefore, to simulate a Johnson X in MATLAB, simply do the following two steps: (a) Z = randn; (b) X can be simulated by inserting Z into Equation (2.54).

2.2.5 Chi-squared PDF model

Suppose that $Z_1, Z_2, \ldots, Z_v$ are independent identically distributed (i.i.d.) standard normal. Consider the following random variable Q:

$$Q = Z_1^2 + Z_2^2 + \ldots + Z_v^2 \tag{2.55}$$

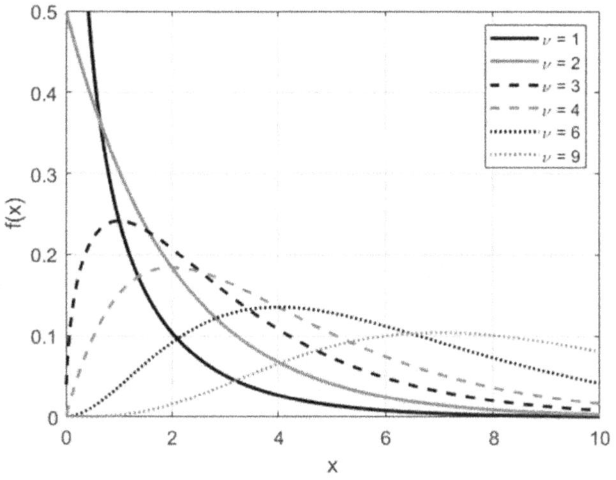

Figure 2.8 Some chi-squared PDFs.

This random variable is called the chi-squared random variable with degree of freedom (DOF) = ν. The chi-squared distribution is denoted by $\chi^2(\nu)$. It has the following PDF:

$$f(q) = \frac{1}{\sqrt{2}^{\nu} \cdot \Gamma(\nu/2)} q^{\nu/2-1} \cdot e^{-q/2} \tag{2.56}$$

where Γ(.) denotes the Gamma function. Figure 2.8 shows some chi-squared PDFs with various DOFs.

2.2.6 Inverse-Gamma PDF model

In Bayesian analysis, it is common to adopt the inverse-Gamma distribution, denoted by IG(α, β), to model an unknown variance (σ^2) because it is the conjugate prior PDF of σ^2 for a normal likelihood function. The concept of a conjugate prior will be elaborated in Section 2.5. The PDF for the inverse-Gamma distribution can be expressed as

$$f(\sigma^2) = \frac{\beta^{\alpha}}{\Gamma(\alpha)} \sigma^{-2(\alpha+1)} \exp\left(\frac{-\beta}{\sigma^2}\right) \tag{2.57}$$

where $\alpha > 0$ is the shape parameter; $\beta > 0$ is the scale parameter. Figure 2.9 shows some inverse-Gamma PDFs with various α and β. When both α and β are small, the PDF for IG(α,β) is flat, that is, the prior PDF for σ^2 is "non-informative" (Baecher and Christian 2003).

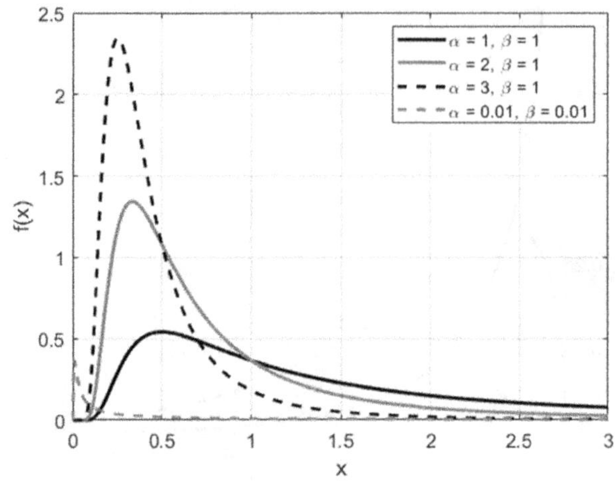

Figure 2.9 Some inverse-Gamma PDFs.

2.2.7 Student's t-distribution model

Student's t-distribution, denoted by t(ν), is a generalization of the standard normal distribution. Its PDF has the following expression:

$$f(x) = \frac{\Gamma\left[(\nu+1)/2\right]}{\sqrt{\nu\pi}\cdot\Gamma(\nu/2)}\left(1+\frac{x^2}{\nu}\right)^{-(\nu+1)/2} \tag{2.58}$$

where ν is the DOF. Figure 2.10 shows some t-distribution PDFs with various DOFs. When ν = ∞, the t-distribution becomes the standard normal

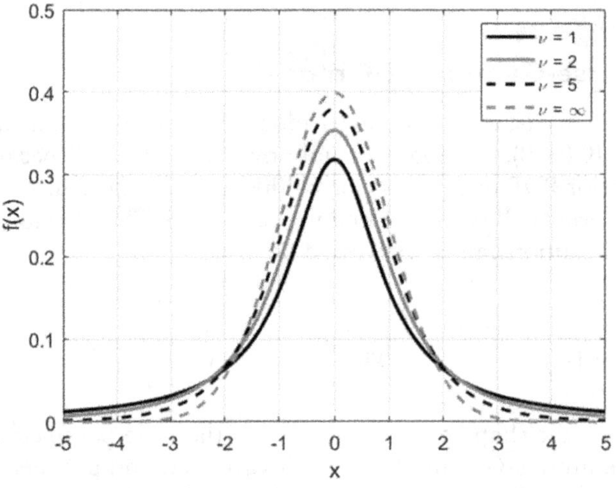

Figure 2.10 Some t-distribution PDFs.

distribution. The non-standardized t-distribution, denoted by $t(\nu,\mu,s^2)$, further generalizes the t-distribution with a location parameter (μ) and scale parameter (s). Its PDF has the following expression:

$$f(x)=\frac{\Gamma[(\nu+1)/2]}{\sqrt{\nu\pi}\cdot s\cdot\Gamma(\nu/2)}\left(1+\frac{1}{\nu}\frac{(x-\mu)^2}{s^2}\right)^{-(\nu+1)/2} \quad (2.59)$$

The non-standardized t-distribution arises in the Bayesian analysis on a normal model $X \sim N(\mu,\sigma^2)$ with an unknown variance (σ^2). Moreover, the unknown σ^2 is assumed to follow $IG(\alpha = \nu/2, \beta = \nu s^2/2)$. It turns out that when the unknown σ^2 is marginalized out by the Total Probability Theorem, X is distributed as $t(\nu,\mu,s^2)$. Namely,

$$\int \frac{1}{\sqrt{2\pi}}\frac{1}{\sigma}\exp\left[\frac{-(x-\mu)^2}{2\cdot\sigma^2}\right]\times\frac{\beta^\alpha}{\Gamma(\alpha)}\sigma^{-2(\alpha+1)}\exp\left(\frac{-\beta}{\sigma^2}\right)\Bigg|_{\alpha=\nu/2,\beta=\nu s^2/2}\times d\sigma^2$$
$$=\frac{\Gamma[(\nu+1)/2]}{\sqrt{\nu\pi}\cdot s\cdot\Gamma(\nu/2)}\left(1+\frac{1}{\nu}\frac{(x-\mu)^2}{s^2}\right)^{-(\nu+1)/2} \quad (2.60)$$

2.3 BIVARIATE PDF MODELS

Recall that the bivariate PDF of (X, Y) is denoted by f(x,y). A bivariate PDF can be visualized as a surface plot (Figure 2.11a) or a contour plot (Figure 2.11b) defined on the x–y space. The probability of (X, Y) is within a subset

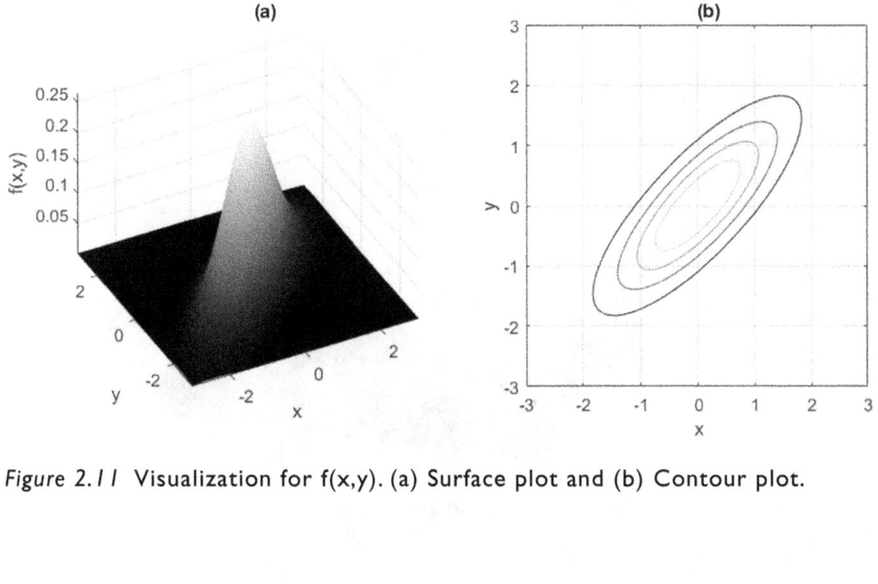

Figure 2.11 Visualization for f(x,y). (a) Surface plot and (b) Contour plot.

Ω in the x–y space can be calculated as the volume under the bivariate PDF within Ω:

$$P(\{X,Y\} \in \Omega) = \iint_{\Omega} f(x,y)dxdy \tag{2.61}$$

Therefore, it is clear that

$$P(-\infty < X < \infty \ \& -\infty < Y < \infty) = \int_{-\infty}^{\infty}\int_{-\infty}^{\infty} f(x,y)dxdy = 1 \tag{2.62}$$

The relationship between the bivariate PDF f(x,y) and the conditional PDF f(x|y) is illustrated in Figure 2.12. Given the information of Y = y_0, a cross section of the f(x,y) can be obtained, which can be written as f(x,y_0). Note that f(x,y_0) is not a PDF of X because its total integral of x is not 1. Its total integral is f(y_0) according to the Marginalization Theorem:

$$\int_{-\infty}^{\infty} f(x,y_0)dx = f(y_0) \tag{2.63}$$

As a result, f(x,y_0)/f(y_0) is a PDF of X because its total integral is 1. Indeed, f(x,y_0)/f(y_0) = f(x|y_0) is a (conditional) PDF of X.

2.3.1 Bivariate normal PDF model

The bivariate normal distribution, denoted by N(μ,C), can be used to model two jointly normally distributed random variables. Let us denote the vector

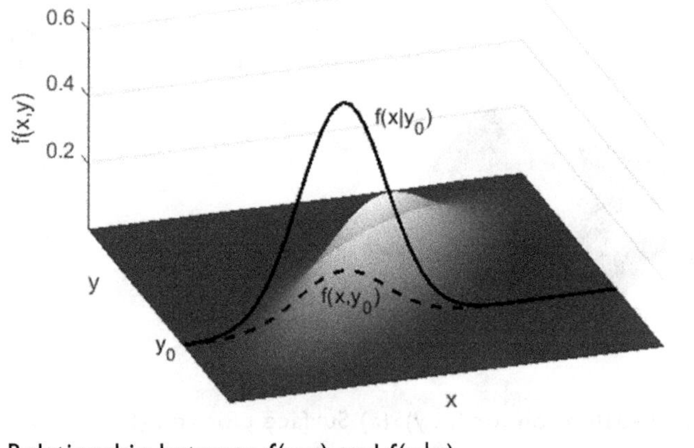

Figure 2.12 Relationship between f(x,y) and f(x|y).

$(X_1\ X_2)^T$ by $\underline{X} \in \mathbf{R}^{2\times1}$. The bivariate PDF of this normal random vector can be described by a mean vector $\underline{\mu} = (\mu_1\ \mu_2)^T \in \mathbf{R}^{2\times1}$ and a covariance matrix $\mathbf{C} \in \mathbf{R}^{2\times2}$:

$$\mathbf{C} = \begin{bmatrix} COV(X_1, X_1) & COV(X_1, X_2) \\ COV(X_2, X_1) & COV(X_2, X_2) \end{bmatrix} = \begin{bmatrix} \sigma_1^2 & \rho_{12}\sigma_1\sigma_2 \\ \rho_{12}\sigma_1\sigma_2 & \sigma_2^2 \end{bmatrix} \tag{2.64}$$

where COV(,) denotes the covariance; σ_i is the standard deviation of X_i; ρ_{ij} is the Pearson product moment correlation coefficient between X_i and X_j:

$$\rho_{ij} = \frac{COV(X_i, X_j)}{\sigma_i \cdot \sigma_j} \tag{2.65}$$

The PDF for the bivariate normal distribution is

$$f(\underline{x}) = \frac{1}{\sqrt{2\pi}^2 \cdot \sqrt{|C|}} \exp\left[\frac{-(\underline{x} - \underline{\mu})^T C^{-1} (\underline{x} - \underline{\mu})}{2}\right] \tag{2.66}$$

where $|\mathbf{C}|$ is the determinant of $\mathbf{C}$ matrix. Figure 2.13 shows the contour plots for two bivariate normal PDFs. Both with $\mu_1 = \mu_2 = 0$ and $\sigma_1 = \sigma_2 = 1$. One is with a positive correlation $\rho_{12} = 0.9$ (Figure 2.13a), whereas the other is with zero correlation $\rho_{12} = 0$ (Figure 2.13b).

One special case is the (independent) bivariate standard normal PDF, that is, $\underline{Z} = [Z_1\ Z_2]^T \sim N(\underline{0}, \mathbf{I}_{2\times2})$, where $\mathbf{I}_{2\times2} \in \mathbf{R}^{2\times2}$ is the identity matrix:

$$f(\underline{z}) = \frac{1}{\sqrt{2\pi}^2} \exp\left(\frac{-\underline{z}^T \times \underline{z}}{2}\right) \tag{2.67}$$

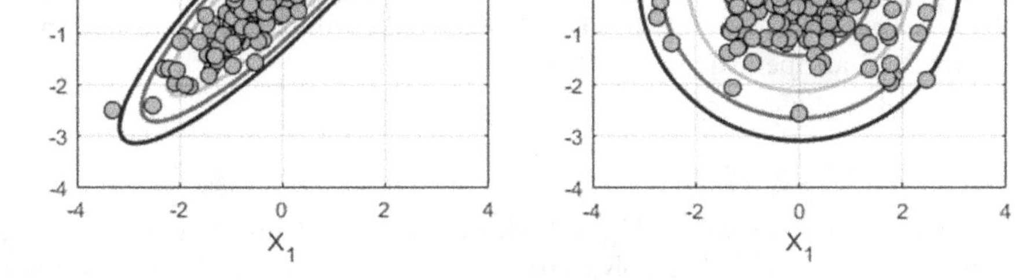

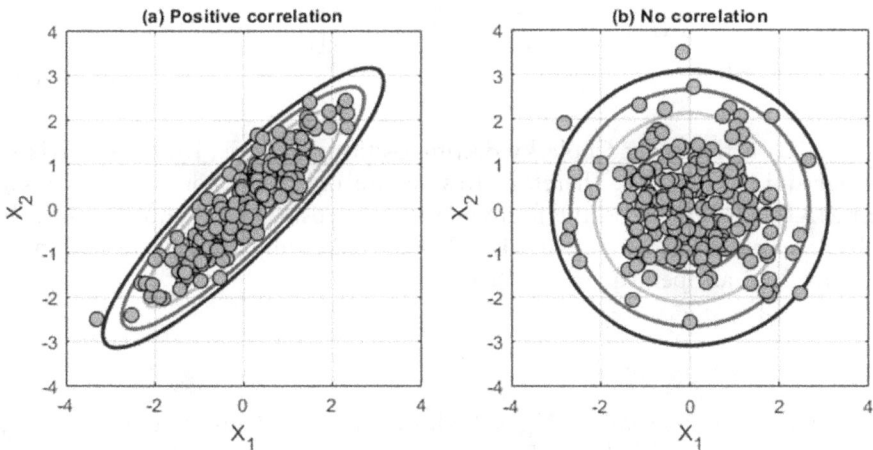

Figure 2.13 Some bivariate normal PDFs. (a) Positive correlation and (b) No correlation.

The MATLAB command for simulating a bivariate standard normal $\underline{Z}$ is simple: $\underline{Z}$ = randn(2,1).

To simulate a bivariate $\underline{X} \sim N(\underline{\mu},\mathbf{C})$, let us first consider simulating $\underline{W} = (W_1,W_2)^T \sim N(\underline{0},\mathbf{R})$, where $\underline{0} = (0,0)^T$, and

$$\mathbf{R} = \begin{bmatrix} 1 & \rho_{12} \\ \rho_{12} & 1 \end{bmatrix} \tag{2.68}$$

Note that $\underline{W}$ has the same correlation matrix $\mathbf{R}$ as $\underline{X}$. The main difference is that $W_i \sim N(0,1)$ is standard normal, but $X_i \sim N(\mu_i,\sigma_i^2)$ is not standard normal. It is clear that

$$\underline{X} = \underline{\mu} + \begin{bmatrix} \sigma_1 & 0 \\ 0 & \sigma_2 \end{bmatrix} \cdot \underline{W} \tag{2.69}$$

It can be shown that $\underline{W}$ can be expressed in terms of two independent standard normal variables $\underline{Z} = (Z_1,Z_2)^T \sim N(\underline{0},\mathbf{I}_{2\times2})$:

$$\underline{W} = \begin{bmatrix} W_1 \\ W_2 \end{bmatrix} = \begin{bmatrix} 1 & 0 \\ \rho_{12} & \sqrt{1-\rho_{12}^2} \end{bmatrix} \begin{bmatrix} Z_1 \\ Z_2 \end{bmatrix} = \mathbf{L}_R \cdot \underline{Z} \tag{2.70}$$

where $\mathbf{L}_R \in \mathbf{R}^{2\times2}$ is the lower Cholesky decomposition of $\mathbf{R}$ (i.e., $\mathbf{L}_R$ is lower triangular and $\mathbf{L}_R \times \mathbf{L}_R^T = \mathbf{R}$). Combining Equations (2.69) and (2.70), we have

$$\begin{aligned} \underline{X} &= \underline{\mu} + \begin{bmatrix} \sigma_1 & 0 \\ 0 & \sigma_2 \end{bmatrix} \cdot \begin{bmatrix} 1 & 0 \\ \rho_{12} & \sqrt{1-\rho_{12}^2} \end{bmatrix} \cdot \underline{Z} = \underline{\mu} + \begin{bmatrix} \sigma_1 & 0 \\ \rho_{12}\sigma_2 & \sigma_2\sqrt{1-\rho_{12}^2} \end{bmatrix} \cdot \\ \underline{Z} &= \underline{\mu} + \mathbf{L}_C \cdot \underline{Z} \end{aligned} \tag{2.71}$$

where $\mathbf{L}_C$ is the lower Cholesky decomposition of $\mathbf{C}$ (i.e., $\mathbf{L}_C$ is lower triangular and $\mathbf{L}_C \times \mathbf{L}_C^T = \mathbf{C}$). Therefore, to simulate a bivariate normal $\underline{X} \sim N(\underline{\mu},\mathbf{C})$ in MATLAB, simply do the following three steps: (a) $\mathbf{L}_C$ = chol(C,'lower'); (b) $\underline{Z}$ = randn(2,1); (c) $\underline{X} = \underline{\mu} + \mathbf{L}_C \times \underline{Z}$. The MCS samples of $\underline{X}$ are shown in Figure 2.13 as open squares.

2.3.2 Bivariate Nataf model

For two random variables (X_1, X_2) whose marginal PDFs are not Gaussian, they cannot be modeled as a bivariate normal PDF. Suppose that the marginal CDFs [F_1(.) and F_2(.)] and quantile functions [F_1^{-1}(.) and F_2^{-1}(.)] are available.

A simple strategy is to first convert (X_1, X_2) to two (correlated) standard normal variables (W_1, W_2) through the following CDF transformation:

$$W_i = \Phi^{-1}\left[F_i(X_i)\right] \tag{2.72}$$

where $F_i(.)$ is the CDF of X_i. It is clear that W_i is distributed as the standard normal distribution N(0,1). For the bivariate Nataf model (Nataf 1962; Liu and Der Kiureghian 1986), $\underline{W}$ = $(W_1, W_2)^T$ are further *assumed* to follow a bivariate normal N($\underline{0}$,**R**), where **R** = [1 ρ_{12}; ρ_{12} 1] is the correlation matrix for $\underline{W}$, and ρ_{12} is the correlation between W_1 and W_2. Note that the fact that W_1 and W_2 individually follow the standard normal distribution does not guarantee that $\underline{W}$ follows a bivariate normal distribution. Under the Nataf model, $\underline{W}$ is *assumed* to follow a bivariate normal distribution. To simulate a bivariate Nataf $\underline{X}$, simply first simulate $\underline{W}$ ~ N($\underline{0}$,**R**). Then, (X_1, X_2) can be obtained by

$$X_i = F_i^{-1}\left[\Phi(W_i)\right] \tag{2.73}$$

2.4 MULTIVARIATE PDF MODELS

2.4.1 Multivariate normal PDF model

The PDF for the multivariate normal distribution, denoted by N($\underline{\mu}$,C), is

$$f(\underline{x}) = \frac{1}{\sqrt{2\pi}^{n} \cdot \sqrt{|C|}} \exp\left[\frac{-(\underline{x}-\underline{\mu})^T C^{-1}(\underline{x}-\underline{\mu})}{2}\right] \tag{2.74}$$

where $\underline{X} = (X_1\ X_2 \ldots X_n)^T \in \mathbf{R}^{n\times1}$ (n is the dimension of $\underline{X}$); $\underline{\mu} = (\mu_1\ \mu_2 \ldots \mu_n)^T \in \mathbf{R}^{n\times1}$; $C \in \mathbf{R}^{n\times n}$ is the covariance matrix:

$$C = \begin{bmatrix} \sigma_1^2 & \rho_{12}\sigma_1\sigma_2 & \cdots & \rho_{1n}\sigma_1\sigma_n \\ & \sigma_2^2 & \cdots & \rho_{2n}\sigma_2\sigma_n \\ & & & \vdots \\ \text{symmetric} & & & \sigma_n^2 \end{bmatrix} \tag{2.75}$$

ρ_{ij} is the Pearson product moment correlation coefficient between X_i and X_j. One special case is the (independent) multivariate standard normal PDF: $\underline{Z}$ = $[Z_1\ Z_2 \ldots Z_n]^T$ ~ N($\underline{0}$,$\mathbf{I}_{n\times n}$):

$$\Phi(\underline{z}) = \frac{1}{\sqrt{2\pi}^{n}} \exp\left(\frac{-\underline{z}^T \times \underline{z}}{2}\right) \tag{2.76}$$

To simulate a multivariate normal $\underline{X} \sim N(\underline{\mu},C)$ in MATLAB, simply do the similar steps for the bivariate one: (a) L_C = chol(C,'lower'); (b) $\underline{Z}$ = randn(n,1); (c) $\underline{X} = \underline{\mu} + L_C \times \underline{Z}$. There is a connection between multivariate (including bivariate) normal random variables and the chi-squared random variable. If $\underline{X} \in R^{n\times1} \sim N(\underline{\mu},C)$, the following random variable (Q) is distributed as the chi-squared distribution with DOF = n:

$$Q = \left(\underline{x} - \underline{\mu}\right)^T C^{-1} \left(\underline{x} - \underline{\mu}\right) \tag{2.77}$$

2.4.2 Multivariate Nataf model

Let $\underline{X} = (X_1\ X_2\ \ldots\ X_n)^T$ and suppose that the marginal CDFs $[F_1(.)\ \ldots\ F_n(.)]$ and quantile functions $[F_1^{-1}(.)\ \ldots\ F_n^{-1}(.)]$ are available. $\underline{X}$ and $\underline{W}$ are related to each other through the CDF transformation:

$$W_i = \Phi^{-1}\left[F_i\left(X_i\right)\right] \tag{2.78}$$

It is clear that W_i is distributed as the standard normal distribution N(0,1). For the multivariate Nataf model, $\underline{W} = (W_1, W_2\ \ldots, W_n)^T$ are further *assumed* to follow a multivariate normal $N(\underline{0},\mathbf{R})$:

$$\mathbf{R} = \begin{bmatrix} 1 & \rho_{12} & \cdots & \rho_{1n} \\ & 1 & \cdots & \rho_{2n} \\ & & & \vdots \\ \text{sym.} & & & 1 \end{bmatrix} \tag{2.79}$$

where ρ_{ij} is the correlation between W_i and W_j. To simulate a multivariate Nataf $\underline{X}$, simply do the following steps: (a) simulate $\underline{W} \sim N(\underline{0},\mathbf{R})$; (b) $(X_1, X_2\ \ldots, X_n)$ can be obtained by $X_i = F_i^{-1}[\Phi(W_i)]$.

2.4.3 Inverse-Wishart PDF model

In Bayesian analysis, it is common to adopt the inverse-Wishart distribution, denoted by $IW(\mathbf{\Sigma},\nu)$, to model an unknown covariance matrix (C) because it is the conjugate prior PDF of C for a multivariate normal likelihood function. The concept of conjugate prior will be elaborated in Section 2.5. The PDF for the inverse-Wishart distribution can be expressed as

$$f(C) = \frac{\left|\Sigma\right|^{\nu/2}}{2^{\nu n/2}\Gamma_n\left(\nu/2\right)}\left|C\right|^{-(\nu+n+1)/2} e^{-0.5\times tr\left(\Sigma\times C^{-1}\right)} \tag{2.80}$$

where $\mathbf{C} \in R^{n\times n}$ is a covariance matrix; $\boldsymbol{\Sigma} \in R^{n\times n}$ is the inverse-Wishart scale matrix (which is positive-definite); ν is the degree of freedom; $\Gamma_n(.)$ is the multivariate Gamma function; tr(.) is the trace of a square matrix (sum of diagonals). The IW PDF is relatively flat (weakly informative) when $\boldsymbol{\Sigma} = \mathbf{I}_{n\times n}$ and $\nu = n + 1$. In MATLAB, the command $\mathbf{C}$ = iwishrnd($\boldsymbol{\Sigma}$,ν) can be used to simulate $\mathbf{C} \sim IW(\boldsymbol{\Sigma},\nu)$.

2.4.4 Wishart PDF model

It is appropriate to adopt the Wishart distribution, denoted by $W(\boldsymbol{\Psi},\lambda)$, to model an unknown inverse-Wishart scale matrix ($\boldsymbol{\Sigma}$) because it is the conjugate prior PDF of $\boldsymbol{\Sigma}$ for an inverse-Wishart likelihood function. The PDF for the Wishart distribution can be expressed as

$$\boldsymbol{\Sigma} \sim W(\boldsymbol{\Sigma};\boldsymbol{\Psi},\lambda) = \frac{|\boldsymbol{\Psi}|^{-\lambda/2}}{2^{\lambda n/2}\Gamma_n(\lambda/2)}|\boldsymbol{\Sigma}|^{(\lambda-n-1)/2} e^{-0.5\times tr(\boldsymbol{\Sigma}\times\boldsymbol{\Psi}^{-1})} \tag{2.81}$$

where $\boldsymbol{\Sigma} \in R^{n\times n}$; $\boldsymbol{\Psi} \in R^{n\times n}$ is the Wishart scale matrix (which is also positive definite); λ is the DOF; the Wishart PDF is relatively flat (weakly informative) when $\boldsymbol{\Psi}$ = a diagonal matrix with large diagonals and $\lambda = n + 2$. In MATLAB, the command $\boldsymbol{\Sigma}$ = wishrnd($\boldsymbol{\Psi}$,λ) can be used to simulate $\boldsymbol{\Sigma} \sim W(\boldsymbol{\Psi},\lambda)$.

2.5 CONJUGATE PRIOR PDFs

In Bayesian analysis, if the posterior PDF $f(\underline{\theta}|\underline{D})$ ($\underline{D}$ denotes the collection of observed data, and $\underline{\theta}$ denotes the collection of the unknown parameters) is in the same PDF type as the prior PDF $f(\underline{\theta})$, the prior PDF is called a conjugate prior for the likelihood function $f(\underline{D}|\underline{\theta})$.

2.5.1 Normal PDF as conjugate prior for normal likelihood with unknown μ

Consider that $\underline{D}$ contains i.i.d. observations $X^{(1)}, X^{(2)}, \ldots, X^{(m)}$ from a univariate normal population $N(\mu,\sigma^2)$, where μ is unknown but σ^2 is known. Note that $f(\underline{D}|\mu)$ is a normal likelihood function as a function of the unknown μ:

$$f(\underline{D} \mid \mu) = \prod_{k=1}^{m} \frac{1}{\sqrt{2\pi}\cdot\sigma} \exp\left[\frac{-\left(X^{(k)}-\mu\right)^2}{2\cdot\sigma^2}\right] \tag{2.82}$$

It turns out that the normal prior PDF $\mu \sim N(\mu_0, \sigma_0^2)$ is a conjugate prior because the posterior PDF of μ is still normal:

$$f(\mu|\underline{D}) \propto f(\underline{D} \mid \mu) \times f(\mu) \propto \left(\prod_{k=1}^{m} \frac{1}{\sqrt{2\pi} \cdot \sigma} \exp\left[\frac{-\left(X^{(k)} - \mu\right)^2}{2 \times \sigma^2} \right] \right)$$
$$\times \left(\frac{1}{\sqrt{2\pi} \times \sigma_0} \exp\left[\frac{-(\mu - \mu_0)^2}{2 \times \sigma_0^2} \right] \right)$$
$$\propto \exp\left\{ -\left[\mu - \left(\mu_0/\sigma_0^2 + \sum_{k=1}^{m} X^{(k)} \Big/ \sigma^2 \right) \right]^2 \Big/ \left[2 \times \left(1/\sigma_0^2 + m/\sigma^2\right)^{-1} \right] \right\}$$
$$= N\left(\left(\frac{\mu_0}{\sigma_0^2} + \sum_{k=1}^{m} X^{(k)} \Big/ \sigma^2 \right) \Big/ \left(\frac{1}{\sigma_0^2} + \frac{m}{\sigma^2} \right), \left(\frac{1}{\sigma_0^2} + \frac{m}{\sigma^2} \right)^{-1} \right) \tag{2.83}$$

2.5.2 Inverse-Gamma as conjugate prior for normal likelihood with unknown σ^2

Consider again that $\underline{D}$ contains i.i.d. observations $X^{(1)}, X^{(2)}, \ldots, X^{(m)}$ from a univariate normal population $N(\mu,\sigma^2)$, where μ is known but σ^2 is unknown. Note that $f(\underline{D}|\sigma^2)$ is a normal likelihood function as a function of the unknown σ^2:

$$f(\underline{D} \mid \sigma^2) = \prod_{k=1}^{m} \frac{1}{\sqrt{2\pi} \cdot \sigma} \exp\left[\frac{-\left(X^{(k)} - \mu\right)^2}{2 \cdot \sigma^2} \right] \tag{2.84}$$

It turns out that the inverse-Gamma prior PDF $\sigma^2 \sim IG(\alpha_0, \beta_0)$ is a conjugate prior because the posterior PDF of σ^2 is still inverse-Gamma:

$$f\left(\sigma^2|\underline{D}\right) \propto \left(\prod_{k=1}^{m} \frac{1}{\sqrt{2\pi} \cdot \sigma} \exp\left[\frac{-\left(X^{(k)} - \mu\right)^2}{2 \times \sigma^2} \right] \right) \times \left[\frac{\beta_0^{\alpha_0}}{\Gamma(\alpha_0)} \sigma^{-2(\alpha_0+1)} \exp\left(\frac{-\beta_0}{\sigma^2} \right) \right]$$
$$\propto \sigma^{-2\left(\alpha_0 + \frac{m}{2} + 1\right)} \exp\left\{ \left[-\beta_0 - \frac{1}{2} \sum_{k=1}^{m} \left(X^{(k)} - \mu\right)^2 \right] \Big/ \sigma^2 \right\}$$
$$= IG\left(\alpha_0 + \frac{m}{2}, \beta_0 + \frac{1}{2} \sum_{k=1}^{m} \left(X^{(k)} - \mu\right)^2 \right) \tag{2.85}$$

2.5.3 Multivariate normal as conjugate prior for multivariate normal likelihood with unknown $\underline{\mu}$

Consider that $\mathbf{D} \in \mathbf{R}^{n\times m}$ contains i.i.d. observations $\underline{X}^{(1)}, \underline{X}^{(2)}, \ldots, \underline{X}^{(m)}$ from a multivariate normal population $N(\underline{\mu},C)$, where $\underline{\mu} \in \mathbf{R}^{n\times 1}$ is unknown but $C \in \mathbf{R}^{n\times n}$ is known. Note that $f(\mathbf{D}|\underline{\mu})$ is a multivariate normal likelihood function as a function of the unknown $\underline{\mu}$:

$$f(\mathbf{D} \mid \underline{\mu}) = \frac{1}{\sqrt{2\pi}^{m\times n} \cdot |C|^{m/2}} \exp\left[-0.5 \times \sum_{k=1}^{m} \left(\underline{X}^{(k)} - \underline{\mu}\right)^{T} C^{-1} \left(\underline{X}^{(k)} - \underline{\mu}\right)\right] \quad (2.86)$$

It turns out that the multivariate normal prior PDF $\underline{\mu} \sim N(\underline{\mu}_0, C_0)$ is a conjugate prior because the posterior PDF of $\underline{\mu}$ is still multivariate normal:

$$\begin{aligned} & f(\underline{\mu} \mid \mathbf{D}) \propto f(\mathbf{D} \mid \underline{\mu}) \times f\left(\underline{\mu}\right) \propto \\ & \exp\left[-0.5 \times \left(\underline{\mu} - \underline{\mu}_0\right)^{T} C_0^{-1} \left(\underline{\mu} - \underline{\mu}_0\right) - 0.5 \times \sum_{k=1}^{m} \left(\underline{X}^{(k)} - \underline{\mu}\right)^{T} C^{-1} \left(\underline{X}^{(k)} - \underline{\mu}\right)\right] \\ & = N\left(\left(C_0^{-1} + mC^{-1}\right)^{-1}\left(C_0^{-1}\underline{\mu}_0 + C^{-1}\sum_{k=1}^{m}\underline{X}^{(k)}\right), \left(C_0^{-1} + mC^{-1}\right)^{-1}\right) \end{aligned} \quad (2.87)$$

2.5.4 Inverse-Wishart as conjugate prior for multivariate normal likelihood with unknown C

Consider again that $\mathbf{D} \in \mathbf{R}^{n\times m}$ contains i.i.d. observations $\underline{X}^{(1)}, \underline{X}^{(2)}, \ldots, \underline{X}^{(m)}$ from a multivariate normal population $N(\underline{\mu},C)$, where $\underline{\mu}$ is known but C is unknown. Note that $f(\mathbf{D}|C)$ is a multivariate normal likelihood function as a function of the unknown C:

$$f(\mathbf{D} \mid C) = \frac{1}{\sqrt{2\pi}^{m\times n} \cdot |C|^{m/2}} \exp\left[-0.5 \times \sum_{k=1}^{m} \left(\underline{X}^{(k)} - \underline{\mu}\right)^{T} C^{-1} \left(\underline{X}^{(k)} - \underline{\mu}\right)\right] \quad (2.88)$$

It turns out that the inverse-Wishart prior PDF $C \sim IW(\boldsymbol{\Sigma}_0, \nu_0)$ is a conjugate prior because the posterior PDF of C is still inverse-Wishart:

$$\begin{aligned} & f(C|\mathbf{D}) \propto f(\mathbf{D} \mid C) \times f(C) \propto \left(\frac{e^{-0.5\times\sum_{k=1}^{m}\left(\underline{X}^{(k)}-\underline{\mu}\right)^{T} C^{-1}\left(\underline{X}^{(k)}-\underline{\mu}\right)}}{\sqrt{2\pi}^{m\times n} \cdot |C|^{m/2}}\right) \\ & \times \left(\frac{|\boldsymbol{\Sigma}|^{\nu_0/2} |C|^{-(\nu_0+n+1)/2} e^{-0.5\times \mathrm{tr}\left(\boldsymbol{\Sigma}_0 \times C^{-1}\right)}}{2^{\nu_0 n/2}\Gamma_n(\nu_0/2)}\right) \\ & \propto |C|^{-(m+\nu_0+n+1)/2} \exp\left[-0.5 \times \mathrm{tr}\left(\boldsymbol{\Sigma}_0 \times C^{-1}\right) - 0.5 \times \sum_{k=1}^{m}\left(\underline{X}^{(k)} - \underline{\mu}\right)^{T} C^{-1}\left(\underline{X}^{(k)} - \underline{\mu}\right)\right] \end{aligned}$$

(2.89)

Note that a scalar is equal to its trace. Also, tr(**A**×**B**) = tr(**B**×**A**). Finally, the trace sign and summation sign are exchangeable. As a result,

$$
\begin{aligned}
& f(\mathbf{C}|\mathbf{D}) \propto |\mathbf{C}|^{-(m+\nu_0+n+1)/2} \\
& \exp\left[-0.5\times \mathrm{tr}\left(\mathbf{\Sigma}_0\times\mathbf{C}^{-1}\right) - 0.5\times\sum_{k=1}^{m}\mathrm{tr}\left[\left(\underline{X}^{(k)}-\underline{\mu}\right)\left(\underline{X}^{(k)}-\underline{\mu}\right)^T\mathbf{C}^{-1}\right]\right] \\
& \propto |\mathbf{C}|^{-(m+\nu_0+n+1)/2}\exp\left[-0.5\times\mathrm{tr}\left(\left\{\mathbf{\Sigma}_0+\sum_{k=1}^{m}\left(\underline{X}^{(k)}-\underline{\mu}\right)\left(\underline{X}^{(k)}-\underline{\mu}\right)^T\right\}\times\mathbf{C}^{-1}\right)\right] \\
& = \mathrm{IW}\left(\mathbf{\Sigma}_0+\sum_{k=1}^{m}\left(\underline{X}^{(k)}-\underline{\mu}\right)\left(\underline{X}^{(k)}-\underline{\mu}\right)^T, m+\nu_0\right)
\end{aligned}
\tag{2.90}
$$

2.5.5 Wishart as conjugate prior for an inverse-Wishart likelihood with unknown $\mathbf{\Sigma}$

Now consider that $\mathbf{C}^{(1)}$, $\mathbf{C}^{(2)}$, …, $\mathbf{C}^{(m)}$ are i.i.d. observations from an inverse-Wishart population IW($\mathbf{\Sigma}$,ν), where $\mathbf{\Sigma}$ is unknown but ν is known. Note that f($\mathbf{C}^{(1)}$, $\mathbf{C}^{(2)}$, …, $\mathbf{C}^{(m)}|\mathbf{\Sigma}$) is an inverse-Wishart likelihood function as a function of the unknown $\mathbf{\Sigma}$:

$$
f(\mathbf{C}^{(1)},\mathbf{C}^{(2)},\ldots,\mathbf{C}^{(m)}\mid\mathbf{\Sigma}) \propto |\mathbf{\Sigma}|^{m\times\nu/2}\times\exp\left[-0.5\times\mathrm{tr}\left(\mathbf{\Sigma}\times\sum_{k=1}^{m}\mathbf{C}^{(k)^{-1}}\right)\right]
\tag{2.91}
$$

It turns out that the Wishart prior PDF $\mathbf{\Sigma}$ ~ W($\mathbf{\Psi}_0$, λ_0) is a conjugate prior because the posterior PDF of **C** is still Wishart:

$$
\begin{aligned}
& f\left(\mathbf{\Sigma}|\mathbf{C}^{(1)},\mathbf{C}^{(2)},\ldots,\mathbf{C}^{(m)}\right) \propto f\left(\mathbf{C}^{(1)},\mathbf{C}^{(2)},\ldots,\mathbf{C}^{(m)}\mid\mathbf{\Sigma}\right)\times f(\mathbf{\Sigma}) \\
& \propto |\mathbf{\Sigma}|^{m\times\nu/2}\times\exp\left[-0.5\times\mathrm{tr}\left(\mathbf{\Sigma}\times\sum_{k=1}^{m}\mathbf{C}^{(k)^{-1}}\right)\right]\times|\mathbf{\Sigma}|^{(\lambda_0-n-1)/2}\exp\left[-0.5\times\mathrm{tr}\left(\mathbf{\Sigma}\times\mathbf{\Psi}_0^{-1}\right)\right] \\
& \propto |\mathbf{\Sigma}|^{(m\times\nu+\lambda_0-n-1)/2}\times\exp\left[-0.5\times\mathrm{tr}\left(\mathbf{\Sigma}\times\left[\mathbf{\Psi}_0^{-1}+\sum_{k=1}^{m}\mathbf{C}^{(k)^{-1}}\right]\right)\right] \\
& = \mathrm{W}\left(\mathbf{\Psi}_0^{-1}+\sum_{k=1}^{m}\mathbf{C}^{(k)^{-1}}, m\times\nu+\lambda_0\right)
\end{aligned}
\tag{2.92}
$$

Table 2.1 summarizes the aforementioned conjugate priors. This table is not an exhaustive compilation of conjugate priors. More conjugate priors can be found in Fink (1997).

Table 2.1 List of conjugate priors and posteriors

Unknown parameter	*Observations*	*Likelihood*	*Conjugate prior PDF*	*Posterior PDF*
μ	$X^{(1)}, \ldots, X^{(m)}$	Univariate normal with known σ^2	$N(\mu_0,\sigma_0^2)$	$N\left(\left(\frac{\mu_0}{\sigma_0^2}+\sum_{k=1}^{m}X^{(k)}/\sigma^2\right)\Big/\left(\frac{1}{\sigma_0^2}+\frac{m}{\sigma^2}\right),\left(\frac{1}{\sigma_0^2}+\frac{m}{\sigma^2}\right)^{-1}\right)$
σ^2	$X^{(1)}, \ldots, X^{(m)}$	Univariate normal with known μ	$IG(\alpha_0,\beta_0)$	$IG\left(\alpha_0+\frac{m}{2},\beta_0+\frac{1}{2}\sum_{k=1}^{m}\left(X^{(k)}-\mu\right)^2\right)$
$\underline{\mu}$	$\underline{X}^{(1)}, \ldots, \underline{X}^{(m)}$	Multivariate normal with known $\mathbf{C}$	$N(\underline{\mu}_0,\mathbf{C}_0)$	$N\left(\left(\mathbf{C}_0^{-1}+m\mathbf{C}^{-1}\right)^{-1}\left(\mathbf{C}_0^{-1}\underline{\mu}_0+\mathbf{C}^{-1}\sum_{k=1}^{m}\underline{X}^{(k)}\right),\left(\mathbf{C}_0^{-1}+m\mathbf{C}^{-1}\right)^{-1}\right)$
$\mathbf{C}$	$\underline{X}^{(1)}, \ldots, \underline{X}^{(m)}$	Multivariate normal with known $\underline{\mu}$	$IW(\mathbf{\Sigma}_0,\nu_0)$	$IW\left(\Sigma_0+\sum_{k=1}^{m}\left(\underline{X}^{(k)}-\underline{\mu}\right)\left(\underline{X}^{(k)}-\underline{\mu}\right)^T,m+\nu_0\right)$
$\mathbf{\Sigma}$	$\mathbf{C}^{(1)}, \ldots, \mathbf{C}^{(m)}$	Inverse-Wishart with known ν	$W(\mathbf{\Psi}_0,\lambda_0)$	$W\left(\Psi_0^{-1}+\sum_{k=1}^{m}\mathbf{C}^{(k)^{-1}},m\times\nu+\lambda_0\right)$

2.6 MODELS FOR SOIL SPATIAL VARIABILITY

The spatial variability of soil (Vanmarcke 1977, 1983) is caused primarily by the natural geologic processes in soil formation. For the vertical (depth) direction, the spatial data can be decomposed into a smoothly varying trend function t(z) and a fluctuating component w(z) as follows (Phoon and Kulhawy 1999a):

$$\xi(z) = t(z) + w(z) \tag{2.93}$$

where ξ stands for the in situ soil property and z = depth. Figure 2.14 shows a schematic for vertical spatial variability.

Let us denote the site investigation data by $\underline{D} = \{\xi(z_1), \xi(z_2), \ldots, \xi(z_m)\}$, where $(z_1, z_2, \ldots, z_m)$ are investigated depths. The trend t(z) is usually modeled as a smoothly varying function of depth, for example, constant trend t(z) = a, linear trend t(z) = a + b×z, and quadratic trend $t(z) = a + b\times z + c\times z^2$, depending on how real data look like. The fluctuating component w(z) is usually modeled as a zero-mean stationary normal random field (Vanmarcke 1977, 1983; Fenton 1999; Fenton and Griffiths 2008). The term "random field" means "random function of space". It is customary to assume that w(z) can be modeled as a zero-mean (wide-sense) stationary normal random field, which means that (a) $w(z) \sim N(0, \sigma^2)$, where the variance σ^2 does not depend on z, and (b) the auto-covariance $COV[w(z_i),w(z_j)]$ only depends on

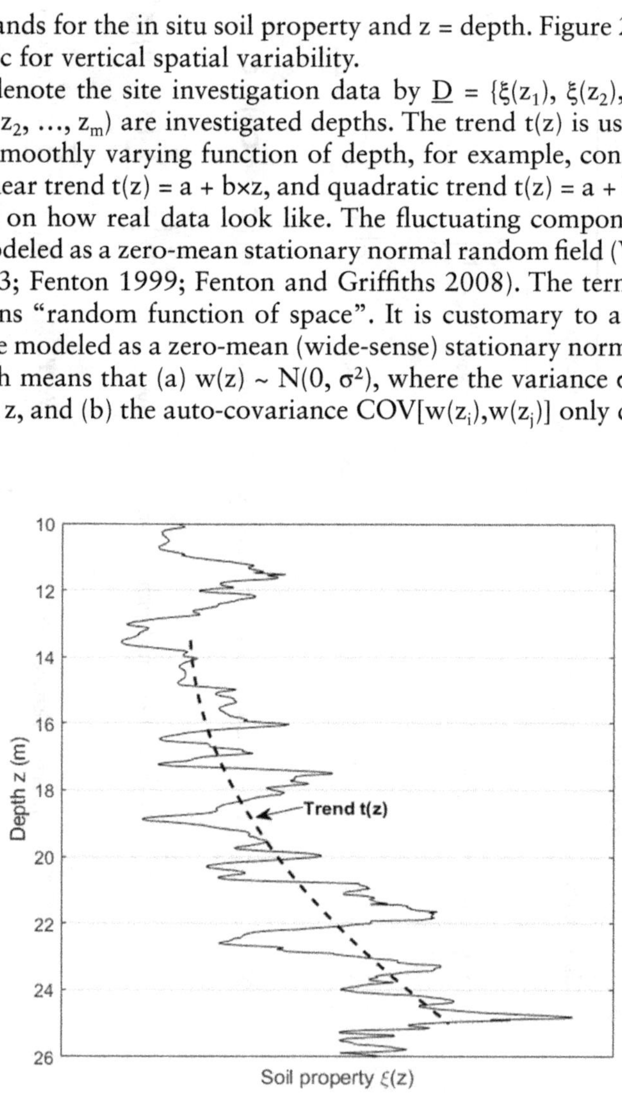

Figure 2.14 Schematic for vertical spatial variability.

the distance (z_i–z_j) but does not depend on the (absolute) location z. The auto-correlation between w(z) and w(z+Δz) can be expressed as

$$\rho\left[w(z),w(z+\Delta z)\right]=\frac{COV\left[w(z),w(z+\Delta z)\right]}{\sqrt{Var\left[w(z)\right]}\cdot\sqrt{Var\left[w(z+\Delta z)\right]}}=\frac{COV\left[w(z),w(z+\Delta z)\right]}{\sigma^2} \tag{2.94}$$

Because the autocovariance COV[w(z),w(z+Δz)] only depends on Δz, ρ[w(z),w(z+Δz)] also only depends on Δz. We call this ρ[w(z),w(z+Δz)] = ρ(Δz) function as the auto-correlation function. As a result, a zero-mean (wide-sense) stationary normal random field w(z) has two types of parameters: (a) the variance of the fluctuation (σ^2); (b) the auto-correlation function ρ(Δz).

The most popular auto-correlation model is the single exponential (SExp) model (Vanmarcke 1977), as shown in Figure 2.15:

$$\rho(\Delta z)=\exp\left(-2\cdot|\Delta z|/\delta\right) \tag{2.95}$$

where δ is called the scale of fluctuation (SOF) (Vanmarcke 1977, 1983; Fenton and Griffiths 2008). It is equal to the total area under the ρ(Δz) function. It can be interpreted as the distance within which w(z) and w(z+Δz) exhibit noticeable correlation (usually positive correlation). There are other auto-correlation models, as shown in Table 2.2. Among them, the

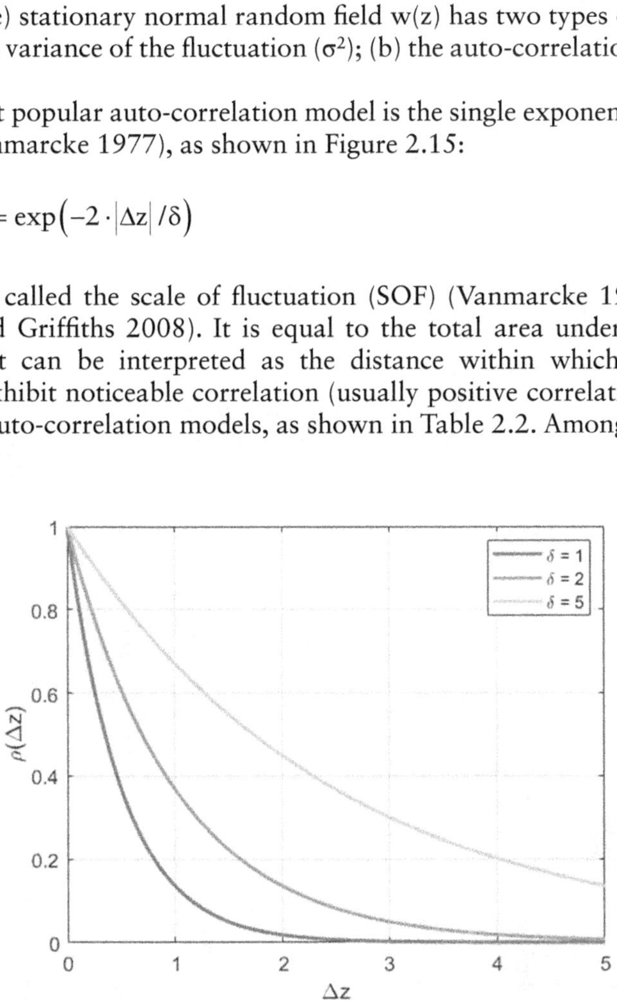

Figure 2.15 Single exponential (SExp) model.

Table 2.2 Some auto-correlation models in geotechnical engineering

Auto-correlation model	*Auto-correlation as a function of lag Δz*	*Smoothness ν*	*Frequency of usage*[a]
Single exponential (SExp)	$\rho(\Delta z) = \exp\left\{\frac{-2\lvert\Delta z\rvert}{\delta}\right\}$	0.5	55%
Second-order Markov (SMK)	$\rho(\Delta z) = \left(1 + 4\frac{\lvert\Delta z\rvert}{\delta}\right)\exp\left\{-4\frac{\lvert\Delta z\rvert}{\delta}\right\}$	1.5	7%
Squared exponential (QExp)	$\rho(\Delta z) = \exp\left\{-\pi\left(\frac{\lvert\Delta z\rvert}{\delta}\right)^2\right\}$	∞ (≈ WM with ν > 3.5)	9%
Spherical (Sph)	$\rho(\Delta z) = \begin{cases} 1 - \frac{9}{8}\left\lvert\frac{\Delta z}{\delta}\right\rvert + \frac{27}{128}\left\lvert\frac{\Delta z}{\delta}\right\rvert^3, & \text{if } \lvert\Delta z\rvert \le \frac{4}{3}\delta; \\ 0, \text{otherwise} \end{cases}$	Outside WM family	15%
Cosine exponential (CosExp)	$\rho(\Delta z) = \exp\left\{-\frac{\lvert\Delta z\rvert}{\delta}\right\}\cos\left\{\frac{\lvert\Delta z\rvert}{\delta}\right\}$	0.5	8%
Binary noise (BN)	$\rho(\Delta z) = \begin{cases} 1 - \lvert\Delta z\rvert/\delta, \text{ if } \lvert\Delta z\rvert \le \delta \\ 0, \text{otherwise} \end{cases}$	Outside WM family	6%
Whittle–Matérn (WM)	$\rho(\Delta z) = \frac{2}{\Gamma(\nu)}\left\{\frac{\sqrt{\pi}\,\Gamma(\nu+0.5)\lvert\Delta z\rvert}{\Gamma(\nu)\delta}\right\}^{\nu} K_{\nu}\left\{\frac{\sqrt{\pi}\,\Gamma(\nu+0.5)\lvert\Delta z\rvert}{\Gamma(\nu)\delta}\right\}$	All ν	New to geotechnical practice

(Modified from Cami et al. 2020).

[a] Updated Phoon et al. (2022) using 297 case studies from ISSMGE-TC304 (2021).

δ = scale of fluctuation; ν = smoothness parameter that reduces the WM model to a specific one-parameter auto-correlation model (e.g., $\nu = 0.5$ produces the Markovian exponential model); Γ = gamma function; and K_ν = modified Bessel function of the second kind with order ν.

Whittle–Matérn (WM) model (Stein 1999; Guttorp and Gneiting 2006) has the following form:

$$\rho(\Delta z) = \frac{2}{\Gamma(\nu)} \cdot \left(\frac{\sqrt{\pi} \cdot \Gamma(\nu + 0.5) \cdot |\Delta z|}{\Gamma(\nu) \cdot \delta} \right)^{\nu} K_{\nu} \left(\frac{2\sqrt{\pi} \cdot \Gamma(\nu + 0.5) \cdot |\Delta z|}{\Gamma(\nu) \cdot \delta} \right) \tag{2.96}$$

where $0 < \nu \leq \infty$ is the smoothness parameter: the sample path of w(z) is $\lceil \nu \rceil - 1$ times differentiable in the mean-square sense ($\lceil \nu \rceil$ denotes the integer ceiling, e.g., $\lceil 3.2 \rceil = 4$); K_ν is the modified Bessel function of the second kind with order ν. For $\nu = 0.5$, 1.5, and ∞, the WM model reduces to the SExp, second-order Markov (SMK), and squared exponential (QExp) models in Table 2.2, respectively.

With the zero-mean stationary normal random field model, $\underline{w} = [w(z_1)\ w(z_2)\ \ldots\ w(z_m)]^T$ follow the multivariate normal PDF $N(\underline{0},\mathbf{C})$:

$$f(\underline{w}) = \frac{1}{\sqrt{2\pi}^{m} \cdot \sqrt{|\mathbf{C}|}} \exp\left(\frac{-\underline{w}^T \mathbf{C}^{-1} \underline{w}}{2} \right) \tag{2.97}$$

where $\mathbf{C} \in \mathbf{R}^{m \times m}$ is the auto-covariance matrix of $\underline{w}$, $\mathbf{C} = \sigma^2 \times \mathbf{R}$; $\mathbf{R} \in \mathbf{R}^{m \times m}$ is the auto-correlation matrix of $\underline{w}$:

$$\mathbf{R} = \begin{bmatrix} 1 & \rho(z_1 - z_2) & \cdots & \rho(z_1 - z_m) \\ & 1 & \cdots & \rho(z_2 - z_m) \\ & & & \vdots \\ \text{symmetric} & & & 1 \end{bmatrix} \tag{2.98}$$

Given the trend function t(z) and the parameters of the stationary normal random field (including variance σ^2 and auto-correlation parameters such as δ), samples of the soil property ξ can be simulated at the depths (z_1, z_2, ..., z_m) by the following procedure: (a) based on the auto-correlation parameters, determine **R**, then compute $\mathbf{C} = \sigma^2 \times \mathbf{R}$; (b) simulate $\underline{w} \sim N(\underline{0},\mathbf{C})$ [MATLAB command: L = chol(C,'lower'); $\underline{w}$ = L×randn(m,1)]; (c) let $\underline{\xi} = \underline{t} + \underline{w}$, where $\underline{\xi} = [\xi(z_1)\ \xi(z_2)\ \ldots\ \xi(z_m)]^T$ and $\underline{t} = [t(z_1)\ t(z_2)\ \ldots\ t(z_m)]^T$.

Now consider a set of spatial data of soil property $\xi(z) = t(z) + w(z)$ simulated at the depth range of 0–20 m with 0.1 m sampling interval $z_1 = 0.1$ m, $z_2 = 0.2$ m, ..., $z_{200} = 20$ m (there are m = 200 depths). The underlying trend is a linear trend:

$$t(z) = 100 + 5 \cdot z \tag{2.99}$$

The spatial variability w(z) is a zero-mean stationary normal random field with standard deviation = σ. The auto-correlation function is the WM model

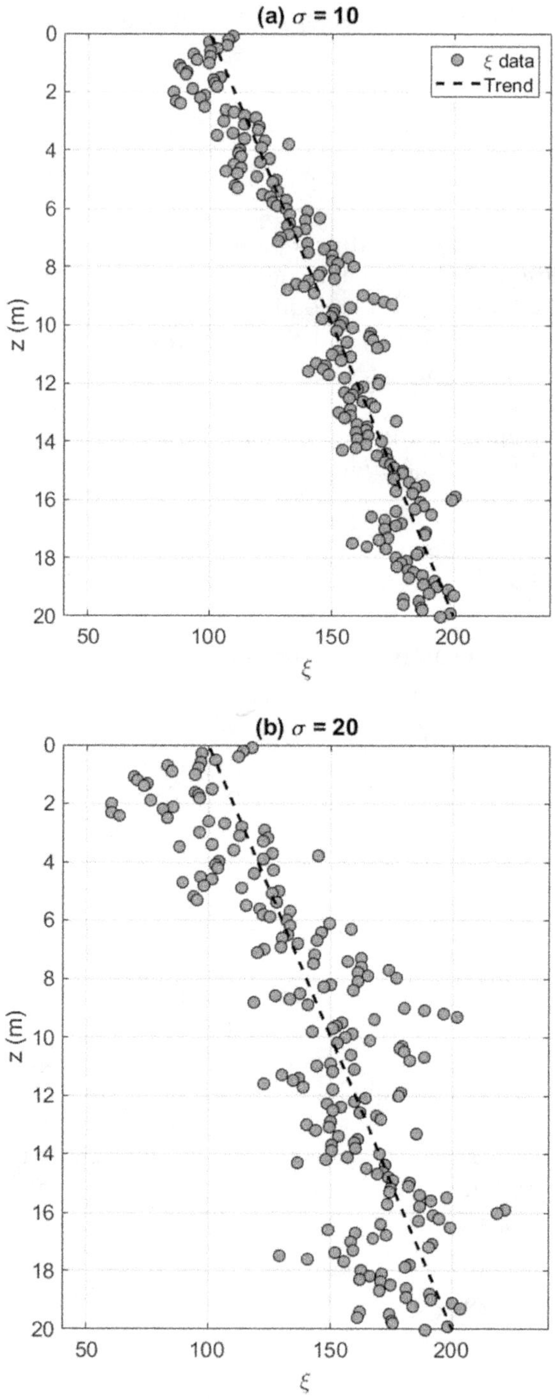

Figure 2.16 Effect of σ on the random field realization.

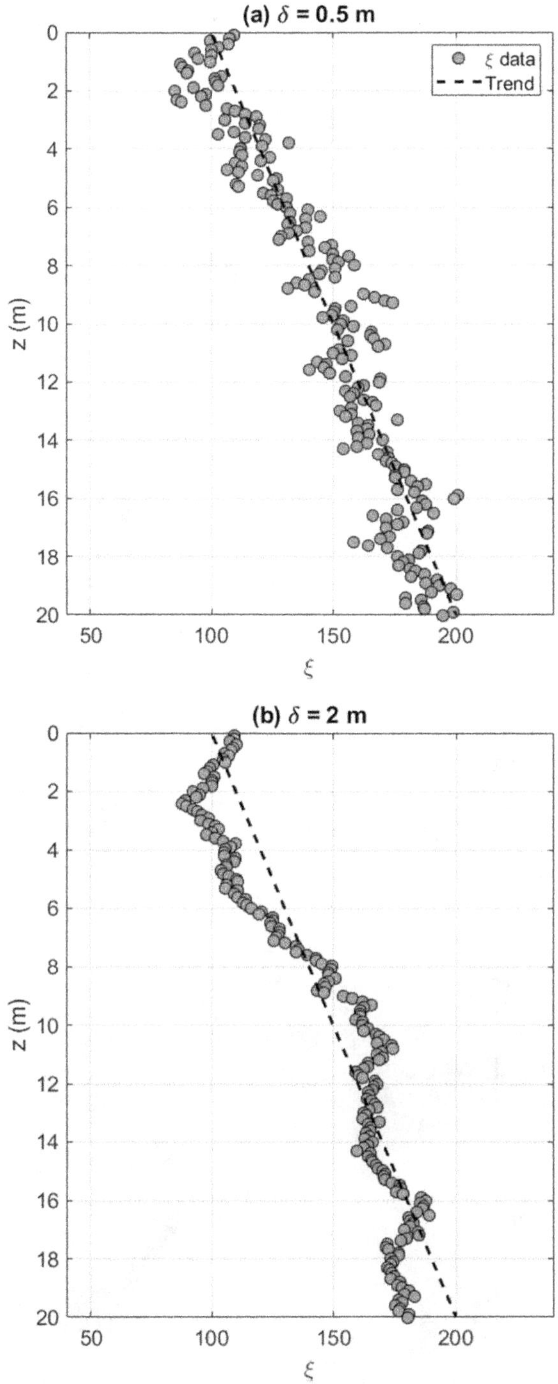

Figure 2.17 Effect of δ in the WM model on the random field realization.

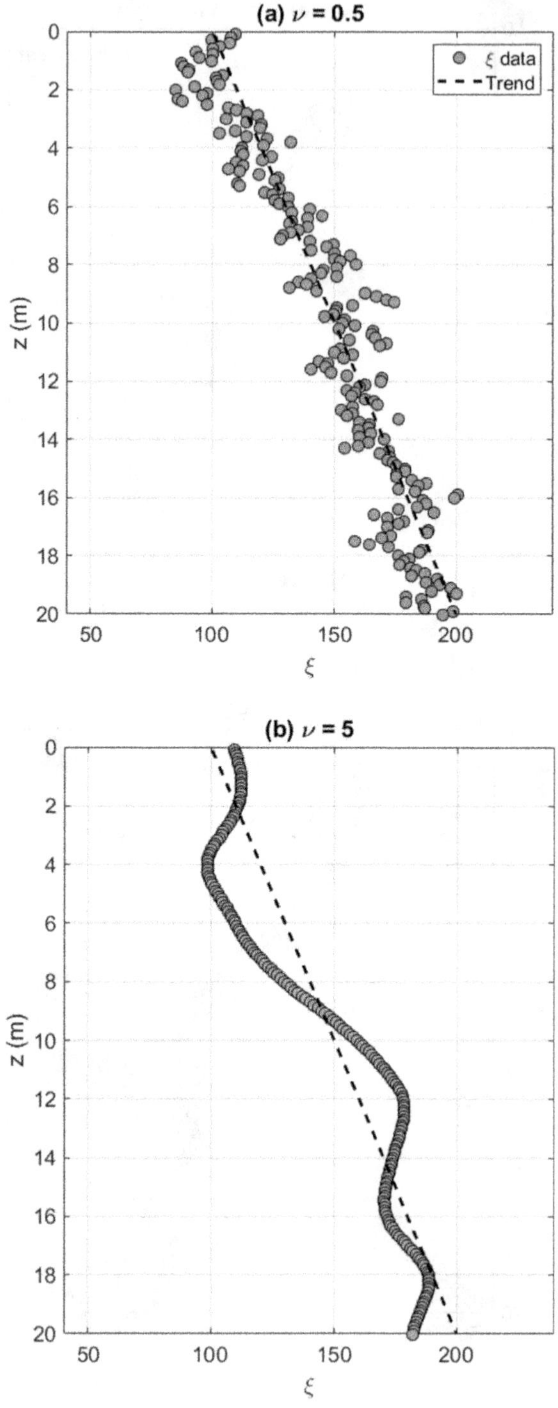

Figure 2.18 Effect of ν in the WM model on the random field realization.

with SOF = δ and smoothness = ν. Figures 2.16–2.18 show the simulated data with respect to depth, $\xi(z_1)$, $\xi(z_2)$, … $\xi(z_m)$. Figure 2.16 shows the effect of σ (Figure 2.16a is for σ = 10, and Figure 2.16b is for σ = 20): σ quantifies the magnitude of fluctuation around the trend. Figure 2.17 shows the effect of δ in the WM model (Figure 2.17a is for δ = 0.5 m, and Figure 2.17b is for δ = 2 m): δ quantifies the correlation length. Figure 2.18 shows the effect of ν in the WM model (Figure 2.18a is for ν = 0.5, and Figure 2.18b is for ν = 5): ν quantifies the smoothness.

Chapter 3

Bayesian parameter estimation and prediction

The purpose of Bayesian parameter estimation is to update the prior probability density function (PDF) for the model parameters (e.g., mean and variance of a normal population) into the posterior PDF with observed data through the likelihood function. The purpose of Bayesian prediction is to further predict future outcomes based on the posterior PDF. In this chapter, Bayesian parameter estimation and prediction procedures are demonstrated using numerical examples. The numerical examples include data simulated from univariate and bivariate normal populations as well as spatially variable data simulated from a stationary normal random field. The Bayesian model selection (Beck and Yuen 2004) will be also briefly demonstrated at the end of this chapter.

3.1 MAXIMUM-LIKELIHOOD METHOD (FREQUENTIST)

The purpose of parameter estimation is to estimate the unknown model parameters given the observed data. The maximum-likelihood (ML) method (Benjamin and Cornell 1970; Baecher and Christian 2003; Ang and Tang 2007; Rossi 2018) is one of the most popular parameter estimation methods. It is also probably the most efficient frequentist method.

EXAMPLE 3.1: Maximum-likelihood estimate with univariate normal data

Assume $X \sim N(\mu, \sigma^2)$, where μ and σ^2 are unknown. Given the independent and identically distributed (i.i.d.) data $\{X^{(1)}, X^{(2)}, \ldots, X^{(m)}\}$ in Table 3.1 (underlying μ = 100 and σ = 20), the purpose is to estimate μ and σ^2. Let us denote the collection of all unknown parameters to be estimated by $\underline{\theta}$. The ML estimates of the

DOI: 10.1201/9781003309765-3

Table 3.1 Generated X data (m = 10) (underlying μ = 100 and σ = 20)

Generated data	*index*
$X^{(k)}$	k
124.06	1
113.35	2
134.75	3
90.52	4
133.54	5
74.40	6
81.15	7
105.41	8
71.43	9
89.81	10

parameters in $\underline{\theta}$ are defined as those that maximize the "likelihood function" $f(\underline{X}|\underline{\theta})$, where $\underline{D} = \{X^{(1)}, X^{(2)}, \ldots, X^{(m)}\}$ denotes the observed data:

$$f(\underline{D} \mid \underline{\theta}) = f(X^{(1)}, X^{(2)}, \ldots, X^{(m)} \mid \mu, \sigma) = \prod_{k=1}^{m} f(X^{(k)} \mid \mu, \sigma)$$
$$= \prod_{k=1}^{m} \frac{1}{\sqrt{2\pi} \cdot \sigma} \exp\left[\frac{-\left(X^{(k)} - \mu\right)^2}{2 \cdot \sigma^2}\right] \quad (3.1)$$

Equivalently, one can maximize the logarithm of the likelihood function:

$$\ln\left[f(X^{(1)}, X^{(2)}, \ldots, X^{(m)} \mid \mu, \sigma)\right] = \sum_{k=1}^{m} \left[-0.5 \ln(2\pi) - \ln(\sigma) - \frac{\left(X^{(k)} - \mu\right)^2}{2 \cdot \sigma^2}\right] \quad (3.2)$$

It is easy to verify that the best estimates are

$$\mu_{MLE} = \frac{1}{m} \sum_{k=1}^{m} X^{(k)} = 101.84 \qquad \sigma_{MLE} = \sqrt{\frac{1}{m} \sum_{k=1}^{m} \left(X^{(k)} - \mu_{MLE}\right)^2} = 22.60 \quad (3.3)$$

Figure 3.1a shows the contour plot for the likelihood function. The cross marker indicates the ML estimates (μ_{MLE}, σ_{MLE}). Figure 3.1b shows the fitted normal PDF with $\mu = \mu_{MLE}$ and $\sigma = \sigma_{MLE}$, plotted together with the observed data. If more X data are generated (m = 100), the likelihood function becomes more peaked (see Figure 3.2a), so the ML estimates become potentially more accurate.

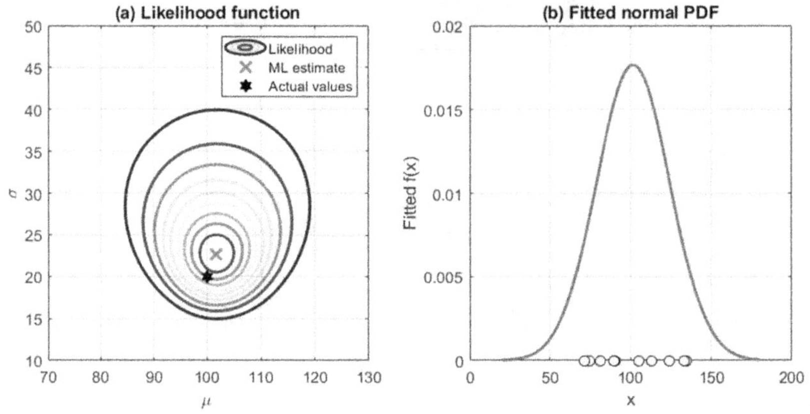

Figure 3.1 Results of ML estimates (m = 10).

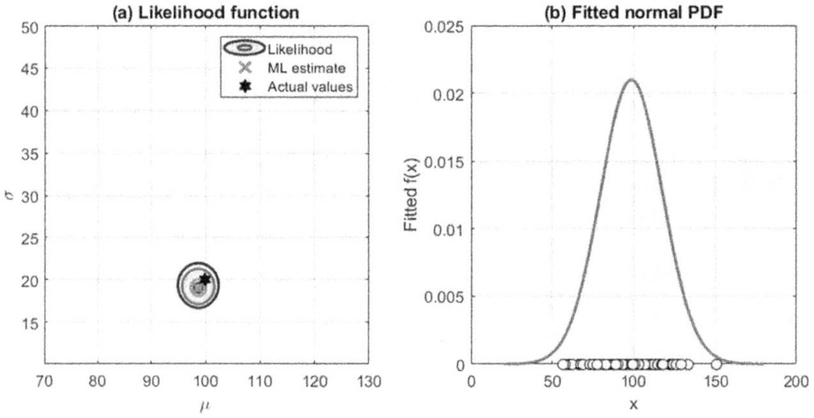

Figure 3.2 Results of ML estimates (m = 100).

3.2 BAYESIAN PROBLEMS WITH UNIVARIATE NORMAL DATA

The ML estimates are frequentist estimates. The main difference between Bayesian and frequentist approaches is in how they view unknown parameters. In the frequentist approach, unknown parameters are considered fixed constants (not random), and point estimates for the parameters can be computed based on measured data. In the Bayesian approach, however, unknown parameters are considered random. The prior PDF is required to quantify their uncertainty. The Bayesian approach interprets probability as the "degree of belief", so the prior PDF reflects the prior degree of belief in the parameter uncertainty. The prior PDF can be updated into the posterior PDF by the measured data via the Bayes theorem (see Equation 1.46):

$$f(x \mid y, a) = \frac{f(y \mid x, a) f(x \mid a)}{\int f(y \mid x, a) f(x \mid a) dx} = \frac{f(y \mid x, a) f(x \mid a)}{f(y \mid a)} \tag{3.4}$$

Recall that proposition **a** is a contextual proposition, that is, **a** = "certain context holds". In this section, this contextual proposition **a** will be omitted, that is, f(x|**a**) will be denoted by f(x), when there is no confusion. Now let $X = \underline{\theta}$ and $Y = \underline{D}$, we have

$$f(x \mid a) = f(\underline{\theta}) = \text{prior PDF of } \underline{\theta} \tag{3.5}$$

$$f(x \mid y, a) = f(\underline{\theta} \mid \underline{D}) = (\text{updated}) \text{ posterior PDF of } \underline{\theta} \tag{3.6}$$

$$f(y \mid x, a) = f(\underline{D} \mid \underline{\theta}) = \text{likelihood function} \tag{3.7}$$

$$f(y \mid a) = f(\underline{D}) = \text{evidence of the assumed context (e.g., the normal PDF model)} \tag{3.8}$$

As a result,

$$f(\underline{\theta} \mid \underline{D}) = \frac{f(\underline{D} \mid \underline{\theta}) f(\underline{\theta})}{f(\underline{D})} \propto f(\underline{D} \mid \underline{\theta}) f(\underline{\theta}) \tag{3.9}$$

or simply (posterior PDF) ∝ (likelihood function) × (prior PDF). The Bayes theorem allows the degree-of-belief probability to be quantified within a consistent mathematical framework. This degree-of-belief probability is not entirely subjective. When more and more data are available, the posterior PDF becomes more and more objective.

3.2.1 Bayesian analysis with conjugate prior

Let us now consider the Bayesian parameter estimation prediction when the prior PDF is conjugate to the likelihood function. Section 2.5 presents several scenarios of conjugate priors. The first example consists of normal data with unknown μ but known σ^2.

EXAMPLE 3.2: Normal data with unknown μ but known σ^2 (Bayesian analysis with conjugate prior)

Assume $X \sim N(\mu, \sigma^2)$, where μ is unknown, but $\sigma = 20$ is known. The prior PDF of $\mu \sim N(\mu_0, \sigma_0^2)$, where μ_0 and σ_0 are specified based on prior information on μ.

BAYESIAN PARAMETER ESTIMATION

Given i.i.d. data $\underline{D} = \{X^{(1)}, X^{(2)}, \ldots, X^{(m)}\}$ in Table 3.1 and the knowledge of $\sigma = 20$, the purpose of the Bayesian parameter estimation is to calculate the posterior PDF of μ. The prior PDF of μ has the following expression:

$$f(\mu) = \frac{1}{\sqrt{2\pi} \cdot \sigma_0} \exp\left[\frac{-(\mu - \mu_0)^2}{2 \cdot \sigma_0^2}\right] \tag{3.10}$$

where μ_0 and σ_0 are specified. The likelihood function of μ has the following expression:

$$f(\underline{D} \mid \mu) = \prod_{k=1}^{m} \frac{1}{\sqrt{2\pi} \cdot \sigma} \exp\left[\frac{-\left(X^{(k)} - \mu\right)^2}{2 \cdot \sigma^2}\right] \tag{3.11}$$

where $\sigma = 20$ is known. The posterior PDF of μ is proportional to the product of the likelihood function of μ and the prior PDF of μ. Note that the normal prior PDF of μ is conjugate to the normal likelihood function of μ (see Section 2.5). According to Table 2.1, the posterior PDF of μ is still normal:

$$f(\mu \mid \underline{D}) = N\left[\left(\frac{\mu_0}{\sigma_0^2} + \sum_{k=1}^{m} X^{(k)} \Big/ \sigma^2\right) \Bigg/ \left(\frac{1}{\sigma_0^2} + \frac{m}{\sigma^2}\right), \left(\frac{1}{\sigma_0^2} + \frac{m}{\sigma^2}\right)^{-1}\right] = N\left(\mu'_0, \sigma'^2_0\right) \tag{3.12}$$

where $\mu'_0 = [\mu_0/\sigma_0^2 + \Sigma X^{(k)}/\sigma^2]/(1/\sigma_0^2 + m/\sigma^2)$ and $\sigma'^2_0 = 1/(1/\sigma_0^2 + m/\sigma^2)$ denote the mean and variance of the posterior PDF of μ. Depending on how the prior parameters (μ_0, σ_0) are specified, the posterior PDF of μ can have roughly the same shape as the likelihood function for a weak prior (e.g., $\mu_0 = 80$, $\sigma_0 = 20$; see Figure 3.3a) and can have a fairly different shape for a strong prior (e.g., $\mu_0 = 80$, $\sigma_0 = 5$; see Figure 3.3b). One can also choose a non-informative (flat) prior by adopting a large σ_0 (e.g., $\mu_0 = 0$, $\sigma_0 = 10{,}000$). In this case, the posterior PDF of μ can have exactly the same shape as the likelihood function.

Because $f(\mu|\underline{D})$ is normal, one can construct the 95% (Bayesian) confidence interval (95% CI) of μ:

$$\mu'_0 - 1.96 \times \sigma'_0 \le \mu \le \mu'_0 + 1.96 \times \sigma'_0 \tag{3.13}$$

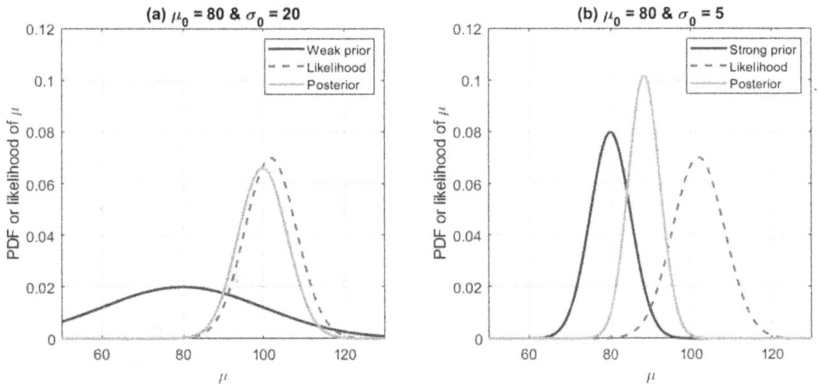

Figure 3.3 Posterior PDF of μ (Example 3.2).

where [−1.96, 1.96] is the 95% CI of the standard normal variable, meaning that there is a 0.95 degree-of-belief probability that the actual μ falls into this 95% CI. Note that the interval should be called a "credible interval" for Bayesian analysis (Jaynes 1976). The term "confidence interval" is in principle a frequentist term. Nonetheless, we will use the term "confidence interval" for both frequentist and Bayesian analysis.

BAYESIAN PREDICTION

Given the data $\underline{D} = \{X^{(1)}, X^{(2)}, \ldots, X^{(m)}\}$ in Table 3.1 and the knowledge of σ = 20, the purpose is to further predict a new i.i.d. X′, that is, $X' \sim N(\mu,\sigma^2)$ (μ is unknown, and σ = 20 is known), but X′ is independent of the data $\underline{D}$. Recall the Total Probability Theorem (Equation 1.45):

$$f(y \mid \mathbf{a}) = \int f(y \mid x, \mathbf{a}) f(x \mid \mathbf{a}) dx \tag{3.14}$$

or simply

$$f(x' \mid \underline{D}) = \int f(x' \mid \mu, \underline{D}) f(\mu \mid \underline{D}) d\mu \tag{3.15}$$

Because X′ is independent of the data $\underline{D}$, the condition on $\underline{D}$ can be dropped:

$$f(x' \mid \mu, \underline{D}) = f(x' \mid \mu) = N\left(\mu, \sigma^2\right) \tag{3.16}$$

As a result,

$$f(x' \mid \underline{D}) = \int \frac{e^{\frac{-(x'-\mu)^2}{2\times\sigma^2}}}{\sqrt{2\pi}\times\sigma} \times \frac{e^{\frac{-(\mu-\mu'_0)^2}{2\times\sigma_0'^2}}}{\sqrt{2\pi}\times\sigma'_0} d\mu = \frac{e^{\frac{-(x'-\mu'_0)^2}{2\times\left(\sigma_0'^2+\sigma^2\right)}}}{\sqrt{2\pi}\times\sqrt{\sigma_0'^2+\sigma^2}} = N\left(\mu'_0, \sigma_0'^2+\sigma^2\right) \tag{3.17}$$

It is remarkable that $f(x'|\underline{D})$ is still normal. One can also construct the 95% CI for X':

$$\mu'_0 - 1.96 \times \sqrt{\sigma'^2_0 + \sigma^2} \leq X' \leq \mu'_0 + 1.96 \times \sqrt{\sigma'^2_0 + \sigma^2} \tag{3.18}$$

Now let us consider the second example consists of normal data with known μ but unknown σ^2.

EXAMPLE 3.3: Normal data with unknown σ^2 but known μ (Bayesian analysis with conjugate prior)

Assume $X \sim N(\mu, \sigma^2)$, where σ is unknown, but $\mu = 100$ is known. Consider the prior PDF of $\sigma^2 \sim IG(\alpha_0,\beta_0)$, where α_0 and β_0 are specified based on prior information on σ^2.

BAYESIAN PARAMETER ESTIMATION

Given i.i.d. data $\underline{D} = \{X^{(1)}, X^{(2)}, \ldots, X^{(m)}\}$ in Table 3.1 and the knowledge of μ = 100, the purpose is to calculate the posterior PDF of σ^2. The prior PDF of σ^2 has the following expression:

$$f(\sigma^2) = \frac{\beta_0^{\alpha_0}}{\Gamma(\alpha_0)} \sigma^{-2(\alpha_0+1)} \exp\left(\frac{-\beta_0}{\sigma^2}\right) \tag{3.19}$$

where α_0 and β_0 are specified. The likelihood function of σ^2 has the following expression:

$$f(\underline{D}|\sigma^2) = \prod_{k=1}^{m} \frac{1}{\sqrt{2\pi} \cdot \sigma} \exp\left[\frac{-\left(X^{(k)} - \mu\right)^2}{2 \cdot \sigma^2}\right] \tag{3.20}$$

where μ = 100 is known. Note that the normal prior PDF of σ^2 is conjugate to the normal likelihood function of σ^2 (see Section 2.5). According to Table 2.1, the posterior PDF of σ^2 is still inverse-Gamma:

$$f(\sigma^2|\underline{D}) = IG\left(\alpha_0 + \frac{m}{2}, \beta_0 + \frac{1}{2}\sum_{k=1}^{m}\left(X^{(k)} - \mu\right)^2\right) = IG(\alpha'_0, \beta'_0) \tag{3.21}$$

where $\alpha'_0 = \alpha_0 + m/2$ and $\beta'_0 = \beta_0 + 0.5 \times \Sigma[X^{(k)}-\mu]^2$ denote the parameters for the posterior IG PDF of σ^2. The 95% CI for σ^2 does not have a simple analytical expression:

$$F_{IG}^{-1}(0.025) \leq \sigma^2 \leq F_{IG}^{-1}(0.975) \tag{3.22}$$

where $F^{-1}_{IG}(p)$ denotes the p-fractile of the posterior IG PDF, which can be calculated by the MATLAB command "1/gaminv(p, α'_0,1/β'_0)".

BAYESIAN PREDICTION

Given the data $\underline{D} = \{X^{(1)}, X^{(2)}, \ldots, X^{(m)}\}$ in Table 3.1 and the knowledge of μ = 100, the purpose is to predict a new i.i.d. X′, that is, $X' \sim N(\mu,\sigma^2)$ (μ = 100 is known, and σ is unknown). According to the Total Probability Theorem:

$$f(x' \mid \underline{D}) = \int f(x' \mid \sigma^2) f(\sigma^2 \mid \underline{D}) d\sigma^2 = \int \frac{e^{\frac{-(x'-\mu)^2}{2\times\sigma^2}}}{\sqrt{2\pi}\times\sigma} \times \frac{\beta_0'^{\alpha'_0}}{\Gamma(\alpha'_0)} \sigma^{-2(\alpha'_0+1)} e^{\frac{-\beta'_0}{\sigma^2}} d\sigma^2 \quad (3.23)$$

Recall that the non-standardized t-distribution $t(\nu,\mu,s)$ arises in the Bayesian analysis on a normal model $X \sim N(\mu,\sigma^2)$ with an unknown variance σ^2 modeled by IG($\alpha = \nu/2$, $\beta = \nu s^2/2$) (see Section 2.2.7). As a result,

$$f(x' \mid \underline{D}) = t\left(\nu = 2\alpha'_0, \mu, s^2 = \beta'_0/\alpha'_0\right) \quad (3.24)$$

which is the non-standardized t-distribution with degree of freedom $\nu = 2\alpha'_0$, location parameter = μ, and scale parameter $s = (\beta'_0/\alpha'_0)^{0.5}$. The 95% CI for X′ does not have a simple analytical expression:

$$F_T^{-1}(0.025) \le X' \le F_T^{-1}(0.975) \quad (3.25)$$

where $F^{-1}_T(p)$ denotes the p-fractile of the non-standardized t-distribution, which can be calculated by the MATLAB command "μ+sqrt(β'_0/α'_0)*tinv(p, 2α'_0)".

3.2.2 Bayesian analysis with non-conjugate prior

The above two examples are with conjugate prior PDFs, which are rare mathematical coincidences in Bayesian analysis. In general, such a coincidence may not happen. The following is an example of non-conjugate priors.

EXAMPLE 3.4: Normal data with unknown σ but known μ (non-conjugate prior)

Assume $X \sim N(\mu, \sigma^2)$, where μ = 100 is known, but σ is unknown. Now consider the lognormal prior PDF of $\sigma \sim LN(m_0, s_0^2)$, where m_0 and s_0 are specified based on prior information on σ.

BAYESIAN PARAMETER ESTIMATION

The prior PDF of σ has the following expression:

$$f(\sigma) = LN(m_0, s_0^2) = \frac{1}{\sqrt{2\pi} \cdot s_0} \frac{1}{\sigma} \exp\left(\frac{-[\ln(\sigma) - m_0]^2}{2 \cdot s_0^2} \right) \tag{3.26}$$

The likelihood function of σ has the following expression:

$$f(\underline{D} \mid \sigma) = \prod_{k=1}^{m} \frac{1}{\sqrt{2\pi} \cdot \sigma} \exp\left[\frac{-\left(X^{(k)} - \mu\right)^2}{2 \cdot \sigma^2} \right] \tag{3.27}$$

The posterior PDF of σ is proportional to the product of the likelihood function of σ and the prior PDF of σ:

$$f(\sigma \mid \underline{D}) \propto \left(\prod_{k=1}^{m} \frac{1}{\sigma} \exp\left[\frac{-\left(X^{(k)} - \mu\right)^2}{2 \cdot \sigma^2} \right] \right) \times \left[\frac{1}{\sigma} \exp\left(\frac{-[\ln(\sigma) - m_0]^2}{2 \cdot s_0^2} \right) \right] \tag{3.28}$$

Note that this posterior PDF is not a lognormal PDF of σ, so the lognormal prior PDF of σ is not conjugate to the normal likelihood function of σ. In fact, there is no analytical expression for this posterior PDF because the normalizing constant [i.e., the evidence = f(D)] has no analytical expression. It is infeasible to express the 95% CI for σ analytically, either.

BAYESIAN PREDICTION

The purpose is to predict an i.i.d. X′:

$$f(x' \mid \underline{D}) = \int f(x' \mid \sigma) f(\sigma \mid \underline{D}) d\sigma \tag{3.29}$$

There is no analytical solution for this integration. It is infeasible to express the 95% CI for X′ analytically, either.

3.2.3 Numerical methods for Bayesian analysis

For Example 3.4, there are no analytical Bayesian solutions. One possible option is to adopt numerical methods for Bayesian analysis: (a) first draw σ samples ~ $f(\sigma|\underline{D})$ for the purpose of parameter estimation; (b) then draw X′ samples ~ $f(x'|\underline{D})$ for the purpose of prediction. The Monte Carlo simulation (MCS) is a method of drawing samples from a PDF, but MCS requires the analytical expression for the quantile function $F^{-1}(.)$. For Example 3.4, $f(\sigma|\underline{D})$ is not a standard PDF, so its CDF does not have analytical expression, not to mention $F^{-1}(.)$. Instead of MCS, one can adopt the Markov chain Monte Carlo (MCMC) methods (Beck and Au 2002; Gilks et al. 1998) to draw samples from $f(\sigma|\underline{D})$. The most well-known MCMC algorithm is the Metropolis–Hastings algorithm (Metropolis et al. 1953; Hastings 1970).

3.2.3.1 *Metropolis–Hastings (MH) algorithm*

Suppose that the target PDF to be sampled is $f(\theta)$:

$$f(\theta) = a \times h(\theta) \tag{3.30}$$

where $h(\theta)$ is a function that can be evaluated; a is some unknown constant. The idea of MH is to create a Markov chain (MC) whose stationary distribution is the same as $f(\theta)$. The procedure of the MH algorithm is as follows.

Procedure for the MH algorithm (with a normal proposal PDF):

1. Initialize $\theta_{t=0}$ = any value ("t" stands for time step).
2. Draw the candidate $\theta_C \sim N(\theta_{t=0}, \sigma_p^2)$ (called the "proposal PDF"; σ_p is chosen by the user).
3. Compute $r = f(\theta_C)/f(\theta_{t=0}) = h(\theta_C)/h(\theta_{t=0})$.
4. Let $\theta_{t=1} = \begin{cases} \theta_C & \text{w.p.} \min(1, r) \\ \theta_{t=0} & \text{otherwise} \end{cases}$ ("w.p." denotes "with probability")
5. Cycle Steps 2–4 to obtain MC samples $\{\theta_t: t = 0, \ldots, T\}$.

It can be shown that $\{\theta_t: t = 0, \ldots, T\}$ are asymptotically (when t is large) distributed as $f(\theta)$.

Note #1: Because the initial state of the Markov chain ($\theta_{t=0}$) is chosen arbitrarily, the chain may need some time to reach its stationary state. We call this period the burn-in period $[0, T_b]$, where T_b denotes the end of the burn-in. The MC samples in the burn-in period may not be distributed as $f(\theta)$ and should be discarded. One way of identifying the burn-in period is to plot the sample value time history (e.g., Figure 3.4) and identify T_b by visual inspection. More rigorous ways of identifying the burn-in period can be found in Gilks et al. (1998).

Note #2: The choice of σ_p can significantly affect the efficiency of the MH algorithm. If σ_p is very small, $r \approx 1$ (no rejection), but adjacent MC samples are highly correlated. If σ_p is very large, r is often small. We get lots of rejections (repeating samples), so the MC samples are also highly correlated. The rule of thumb is to take σ_p to be of the same order as that of the standard deviation of $f(\theta)$. Usually, the performance is the best when the rejection rate is around 50%.

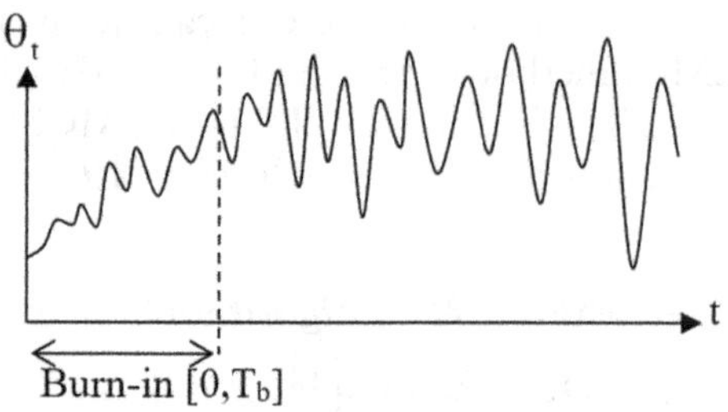

Figure 3.4 Visual inspection of the burn-in period.

EXAMPLE 3.5: Normal data with known μ but unknown σ (Example 3.4 revisited with the MH algorithm)

BAYESIAN PARAMETER ESTIMATION

Recall from Example 3.4 that

$$f(\sigma \mid \underline{D}) = a \times \left(\prod_{k=1}^{m} \frac{1}{\sigma} \exp\left[\frac{-\left(X^{(k)} - \mu\right)^2}{2 \cdot \sigma^2} \right] \right) \times \left[\frac{1}{\sigma} \exp\left(\frac{-\left[\ln(\sigma) - m_0\right]^2}{2 \cdot s_0^2} \right) \right]$$
$$= a \times h(\sigma) \tag{3.31}$$

where a is a constant that does not depend on σ. The parameters for the lognormal prior PDF of σ, LN(m_0, s_0^2), are chosen as m_0 = 3 and s_0 = 1. This prior PDF is shown in Figure 3.5 as the "lognormal prior". Note that h(σ) in Equation

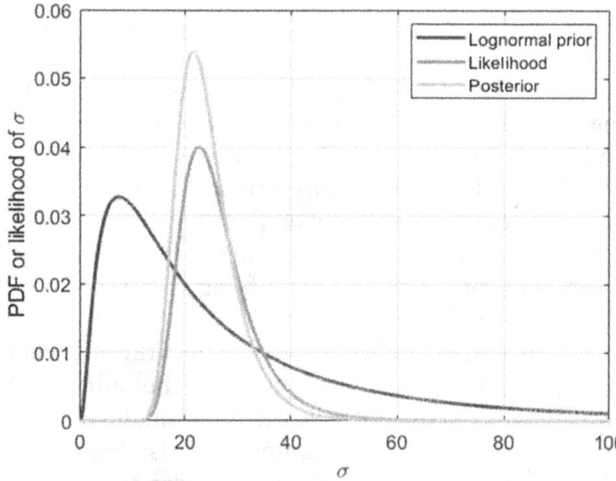

Figure 3.5 Prior, likelihood function, and posterior PDF of σ (Example 3.5).

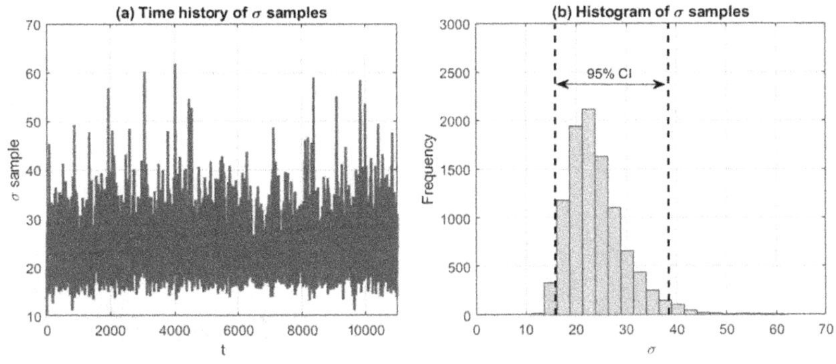

Figure 3.6 Results of the MH algorithm (Example 3.5, $\sigma_p = 10$).

(3.31) can be evaluated because $X^{(1)}, \ldots, X^{(m)}$, m_0, and s_0 are all known numbers. Now consider the MH algorithm with $T = 11000$, $T_b = 1000$, and $\sigma_p = 10$. The σ sample time history is shown in Figure 3.6a. The histogram for the samples after burn-in $\{\sigma_t: t = T_b+1, \ldots, T\}$ is shown in Figure 3.6b. The 95% CI of σ can be estimated as the interval between the sample 0.025-fractile (the lower 2.5% sample) and sample 0.975-fractile (the higher 2.5% sample) (see the vertical dashed lines in Figure 3.6b). If σ_p is changed to 0.1, the σ sample time history is shown in Figure 3.7a. The rejection ratio is very low, the σ sample moves slowly, and the correlation among samples is high. The Markov chain needs a large T to converge to the stationary state, so the histogram in Figure 3.7b is a poor representation of the posterior PDF $f(\sigma|\underline{D})$.

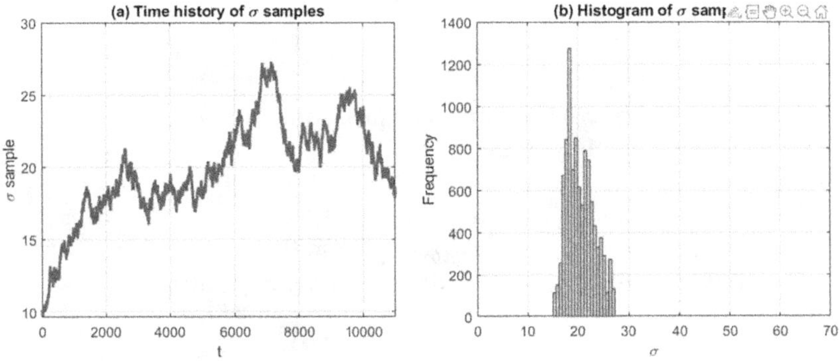

Figure 3.7 Results of the MH algorithm (Example 3.5, $\sigma_p = 0.1$).

BAYESIAN PREDICTION

The purpose is to draw new i.i.d. X′ samples from f(x′|$\underline{D}$). According to the Total Probability Theorem:

$$f(x'|\underline{D}) = \int f(x'|\sigma,\underline{D})f(\sigma|\underline{D})d\sigma = \int f(x'|\sigma)f(\sigma|\underline{D})d\sigma \tag{3.32}$$

According to the Law of Large Numbers:

$$f(x'|\underline{D}) \approx \frac{1}{T-T_b}\sum_{t=T_b+1}^{T} f(x'|\sigma_t) = \frac{1}{T-T_b}\sum_{t=T_b+1}^{T}\frac{1}{\sqrt{2\pi}\cdot\sigma_t}\exp\left[\frac{-(x'-\mu)^2}{2\cdot\sigma_t^2}\right] \tag{3.33}$$

where $\{\sigma_t: t = T_b+1, \ldots, T\}$ are the σ samples ~ f(σ|$\underline{D}$) drawn by the MH algorithm. According to Equation (3.33), the posterior PDF f(x′|$\underline{D}$) can be approximated as a mixture of normal PDFs of X′. As a result, the X′ samples ~ f(x′|$\underline{D}$) can be obtained using the following procedure:

1. Randomly draw σ sample from $\{\sigma_t: t = T_b+1, \ldots, T\}$, and let us denote this sample by $\hat{\sigma}$.
2. Draw a sample $X' \sim N(\mu, \hat{\sigma}^2)$.
3. Cycle Steps 1 and 2 to obtain the desirable number of X′ sample.

The 95% CI of X′ can be estimated as the interval between the sample 0.025-fractile and sample 0.975-fractile.

The MH algorithm can also be generalized to sample a high-dimensional PDF. Suppose the target PDF to be sampled is $f(\theta_1,\theta_2)$:

$$f(\theta_1,\theta_2) = a\cdot h(\theta_1,\theta_2) \tag{3.34}$$

where a is some unknown constant. The procedure of the MH algorithm for a high-dimensional PDF is similar:

1. Initialize $\theta_{1,t=0}$ and $\theta_{2,t=0}$ as any values.
2. Let the candidate $\theta_{1,C} \sim N(\theta_{1,t=0}, \sigma_{p,1}^2)$ and $\theta_{2,C} \sim N(\theta_{2,t=0}, \sigma_{p,2}^2)$.
3. Compute $r = f(\theta_{1,C},\theta_{2,C})/f(\theta_{1,t=0},\theta_{2,t=0}) = h(\theta_{1,C},\theta_{2,C})/h(\theta_{1,t=0},\theta_{2,t=0})$.
4. Let $(\theta_{1,t=1},\theta_{2,t=1}) = \begin{cases} (\theta_{1,C},\theta_{2,C}) & \text{w.p.}\min(1,r) \\ (\theta_{1,t=0},\theta_{2,t=0}) & \text{otherwise} \end{cases}$
5. Cycle Steps 2–4 to obtain MC samples $\{(\theta_{1,t},\theta_{2,t}): t = 0, \ldots, T\}$.

It can be shown that $\{(\theta_{1,t},\theta_{2,t}): t = 0, \ldots, T\}$ are asymptotically (when t is large) distributed as $f(\theta_1,\theta_2)$.

EXAMPLE 3.6: Normal data with unknown μ and σ

Assume $X \sim N(\mu, \sigma^2)$, where μ and σ are both unknown. The prior PDF of $\mu \sim N(\mu_0, \sigma_0^2)$, and prior PDF of $\sigma \sim LN(\sigma; m_0, s_0^2)$, where $\mu_0 = 80$, $\sigma_0 = 20$, $m_0 = 3$, and $s_0 = 1$. Let us further suppose μ and σ are independent in their prior.

BAYESIAN PARAMETER ESTIMATION

Given i.i.d. data $\underline{D} = \{X^{(1)}, X^{(2)}, \ldots, X^{(m)}\}$ in Table 3.1, the purpose is to draw $\underline{\theta} = \{\mu, \sigma\}$ samples from $f(\underline{\theta}|\underline{D})$. Note that

$$
\begin{aligned}
f(\underline{\theta} \mid \underline{D}) &= a \times f(\underline{D} \mid \underline{\theta}) \times f(\underline{\theta}) = a \times f(\underline{D} \mid \mu, \sigma) \times f(\mu) \times f(\sigma) \\
&= a \times \left(\prod_{k=1}^{m} \frac{1}{\sqrt{2\pi} \cdot \sigma} e^{\frac{-\left(x^{(k)} - \mu\right)^2}{2 \times \sigma^2}} \right) \times \left[\frac{1}{\sqrt{2\pi} \cdot \sigma_0} \frac{1}{\sigma} e^{\frac{-(\mu - \mu_0)}{2 \times \sigma_0^2}} \right] \times \left[\frac{1}{\sqrt{2\pi} \cdot s_0} \frac{1}{\sigma} e^{\frac{-\left[\ln(\sigma) - m_0\right]^2}{2 \times s_0^2}} \right] \\
&= a \times h(\mu, \sigma)
\end{aligned}
\tag{3.35}
$$

Using the MH algorithm with $T = 11{,}000$, $T_b = 1000$, $\sigma_{p,\mu} = 10$, and $\sigma_{p,\sigma} = 10$, we can get the posterior samples of $\{(\mu_t, \sigma_t): t = T_b+1, \ldots, T\} \sim f(\underline{\theta}|\underline{D})$. The 95% confidence region (95% CR) for the actual values of $[\mu, \ln(\sigma)]$ (σ is taken logarithm because it cannot be negative) can be constructed based on the $[\mu_t, \ln(\sigma_t)]$ samples by assuming the posterior PDF of $[\mu, \ln(\sigma)]$ to be bivariate normal. This 95% CR is a two-dimensional (2D) ellipse in the μ-$\ln(\sigma)$ space, defined as

$$
\begin{bmatrix} \mu - \mu_{est} \\ \ln(\sigma) - \ln(\sigma)_{est} \end{bmatrix}^T \times \mathbf{C}_{est}^{-1} \times \begin{bmatrix} \mu - \mu_{est} \\ \ln(\sigma) - \ln(\sigma)_{est} \end{bmatrix} \le 5.99
\tag{3.36}
$$

where μ_{est} and $\ln(\sigma)_{est}$ denote the sample means of the μ and $\ln(\sigma)$ samples obtained by the MH algorithm; $\mathbf{C}_{est} \in \mathbf{R}^{2\times2}$ is the sample covariance matrix of the μ and $\ln(\sigma)$ samples; 5.99 is the 0.95-fractile of the chi-squared random variable with $n = 2$ degrees of freedom [computed as chi2inv(0.95,n) in MATLAB]. This 95% CR is shown in Figure 3.8. If more X data are generated ($m = 100$ data points), the likelihood function becomes more peaked. The posterior PDF also becomes more peaked because it is proportional to the likelihood, so the posterior samples become more concentrated, as shown in Figure 3.9.

BAYESIAN PREDICTION

Given i.i.d. data $\underline{D} = \{X^{(1)}, X^{(2)}, \ldots, X^{(m)}\}$, the purpose is to draw $X' \sim f(x'|\underline{D})$. According to the Total Probability Theorem and the Law of Large Numbers:

$$
f(x' \mid \underline{D}) \approx \frac{1}{T - T_b} \sum_{t=T_b+1}^{T} \frac{1}{\sqrt{2\pi} \cdot \sigma_t} \exp\left[\frac{-\left(x' - \mu_t\right)^2}{2 \cdot \sigma_t^2} \right]
\tag{3.37}
$$

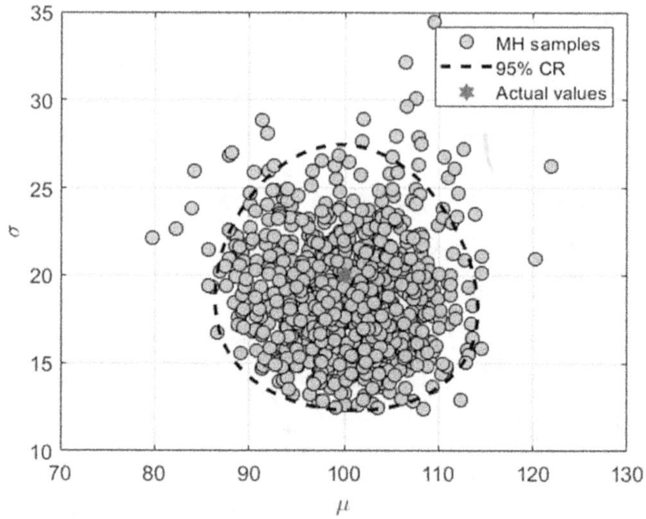

Figure 3.8 Posterior samples of μ and σ and the 95% confidence region (Example 3.6, m = 10).

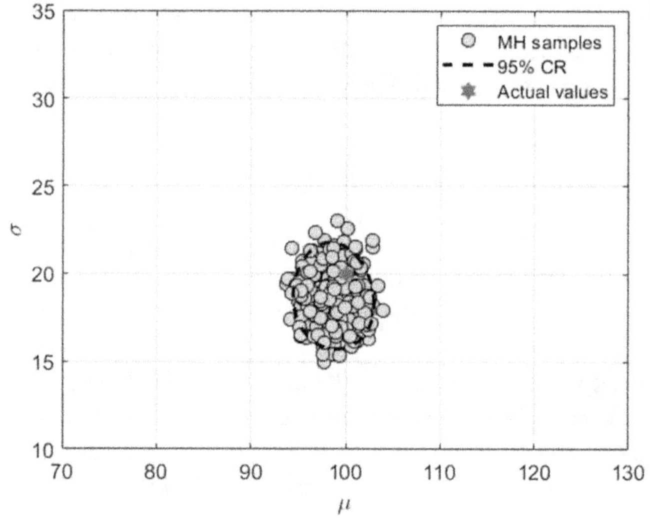

Figure 3.9 Posterior samples of μ and σ and the 95% confidence region (Example 3.6, m = 100).

where $\{(\mu_t,\sigma_t): t = T_b+1, \ldots, T\} \sim f(\underline{\theta}|\underline{D})$ are the samples drawn by the MH algorithm. The X′ samples $\sim f(x'|\underline{D})$ can be obtained using the following procedure:

1. Randomly draw (μ,σ) from $\{(\mu_t,\sigma_t): t = T_b+1, \ldots, T\}$ and denote this sample by $(\hat{\mu},\hat{\sigma})$.
2. Draw a sample $X' \sim N(\hat{\mu},\hat{\sigma}^2)$.
3. Cycle Steps 1 and 2 to obtain the desirable number of X′ sample.

The 95% CI of X′ can be estimated as the interval between the sample 0.025-fractile and sample 0.975-fractile.

3.2.3.2 *Gibbs sampler (GS) algorithm*

The Gibbs sampler (GS) algorithm (Geman and Geman 1984) is another variant of MCMC methods. Suppose that the target PDF to be sampled is $f(\boldsymbol{\Theta}) = f(\underline{\theta}^{group1},\underline{\theta}^{group2},\ldots,\underline{\theta}^{group\ M})$, where $\boldsymbol{\Theta}$ contains all unknown parameters, and $\boldsymbol{\Theta}$ is divided into M groups: $\boldsymbol{\Theta} = \{\underline{\theta}^{group1},\underline{\theta}^{group2},\ldots,\underline{\theta}^{group\ M}\}$. The basic idea is to sample from one group $\underline{\theta}^{group\ i}$ by conditioning on all other groups $\boldsymbol{\Theta}\backslash\underline{\theta}^{group\ i} = \{\underline{\theta}^{group1},\ldots,\underline{\theta}^{group\ i-1},\underline{\theta}^{group\ i+1},\ldots,\underline{\theta}^{group\ M}\}$. This PDF $f(\underline{\theta}^{group\ i}|\boldsymbol{\Theta}\backslash\underline{\theta}^{group\ i})$ is called the fully conditional PDF. The procedure for the GS algorithm is as follows:

1. Initialize $\boldsymbol{\Theta}_{t=0} = \left\{\underline{\theta}_{t=0}^{group\,1},\underline{\theta}_{t=0}^{group\,2},\ldots,\underline{\theta}_{t=0}^{group\,M}\right\}$ at any value.
2. Draw sample $\underline{\theta}_{t=1}^{group\,1} \sim f\left(\underline{\theta}^{group\,1} \mid \underline{\theta}_{t=0}^{group\,2},\ldots,\underline{\theta}_{t=0}^{group\,M}\right)$
3. Draw sample $\underline{\theta}_{t=1}^{group\,2} \sim f\left(\underline{\theta}^{group\,2} \mid \underline{\theta}_{t=1}^{group\,1},\underline{\theta}_{t=0}^{group\,3},\ldots,\underline{\theta}_{t=0}^{group\,M}\right)$
4. Draw sample $\underline{\theta}_{t=1}^{group\,3} \sim f\left(\underline{\theta}^{group\,3} \mid \underline{\theta}_{t=1}^{group\,1},\underline{\theta}_{t=1}^{group\,2},\underline{\theta}_{t=0}^{group\,4},\ldots,\underline{\theta}_{t=0}^{group\,M}\right)$
5. …
6. Draw sample $\underline{\theta}_{t=1}^{group\,M} \sim f\left(\underline{\theta}^{group\,M} \mid \underline{\theta}_{t=1}^{group\,1},\underline{\theta}_{t=1}^{group\,2},\ldots,\underline{\theta}_{t=1}^{group\,M-1}\right)$
7. Cycle Steps 2–6 to obtain MC samples $\{\boldsymbol{\Theta}_t: t = 0, \ldots, T\}$.

It can be shown that these samples will be asymptotically (when t is large) distributed as $f(\boldsymbol{\Theta})$. The samples within the burn-in period are also discarded. Note that there are no proposal PDFs needed during the GS algorithm (no need to specify σ_p, either). This is a major advantage compared to the MH algorithm. There is also no rejection of samples. The GS algorithm is useful when all fully conditional PDFs $f(\underline{\theta}^{group\ i}|\boldsymbol{\Theta}\backslash\underline{\theta}^{group\ i})$ have analytical expressions by adopting conjugate priors.

EXAMPLE 3.7: Normal data with unknown μ and σ (with conjugate priors)

Assume $X \sim N(\mu, \sigma^2)$, where μ and σ are both unknown. The prior PDF of $\mu \sim N(\mu_0, \sigma_0^2)$, and prior PDF of $\sigma^2 \sim IG(\alpha_0,\beta_0)$. Note that these priors are conjugate priors. Furthermore, let us assume non-informative (flat) priors by adopting $\mu_0 = 0$, $\sigma_0 = 10{,}000$, $\alpha_0 = 0.01$, and $\beta_0 = 0.01$. Let us further assume that μ and σ^2 are independent in their priors. Let us divide the unknown variables $\underline{\theta} = \{\mu, \sigma^2\}$ into two groups: μ and σ^2. According to Table 2.1, the fully conditional PDF of μ is still normal and the fully conditional PDF of σ^2 is still inverse-Gamma:

$$f(\mu \mid \sigma, \underline{D}) = N\left[\left(\frac{\mu_0}{\sigma_0^2} + \sum_{k=1}^{m} X^{(k)} \Big/ \sigma^2\right) / \left(\frac{1}{\sigma_0^2} + \frac{m}{\sigma^2}\right), \left(\frac{1}{\sigma_0^2} + \frac{m}{\sigma^2}\right)^{-1}\right] \tag{3.38}$$

$$f(\sigma^2 \mid \mu, \underline{D}) = IG\left(\alpha_0 + \frac{m}{2}, \beta_0 + \frac{1}{2}\sum_{k=1}^{m}\left(X^{(k)} - \mu\right)^2\right) \tag{3.39}$$

BAYESIAN PARAMETER ESTIMATION

Given i.i.d. data $\underline{D} = \{X^{(1)}, X^{(2)}, \ldots, X^{(m)}\}$ in Table 3.1, the purpose is to draw $\underline{\theta}$ samples $\sim f(\underline{\theta}|\underline{D})$. Due to the conjugate priors, the full conditional PDFs have analytical expressions as shown in Eqs. (3.38) and (3.39), so the following GS algorithm can be adopted:

1. Initialize $(\mu_{t=0},\sigma_{t=0})$ at any values.
2. Draw sample $\mu_{t=1} \sim N\left[\left(\frac{\mu_0}{\sigma_0^2} + \sum_{k=1}^{m} X^{(k)} \Big/ \sigma_{t=0}^2\right) \Big/ \left(\frac{1}{\sigma_0^2} + \frac{m}{\sigma_{t=0}^2}\right), \left(\frac{1}{\sigma_0^2} + \frac{m}{\sigma_{t=0}^2}\right)^{-1}\right]$
3. Draw sample $\sigma_{t=1}^2 \sim IG\left(\alpha_0 + \frac{m}{2}, \beta_0 + \frac{1}{2}\sum_{k=1}^{m}\left(X^{(k)} - \mu_{t=1}\right)^2\right)$
4. Cycle Steps 2 and 3 to obtain MC samples $\{(\mu_t,\sigma_t^2): t = 0, \ldots, T\}$.

Using the GS algorithm with $T = 11{,}000$ and $T_b = 1000$, we can get the posterior samples of $\{(\mu_t,\sigma_t): t = T_b+1, \ldots, T\} \sim f(\underline{\theta}|\underline{D})$. The posterior samples as well as the 95% CR are shown in Figure 3.10.

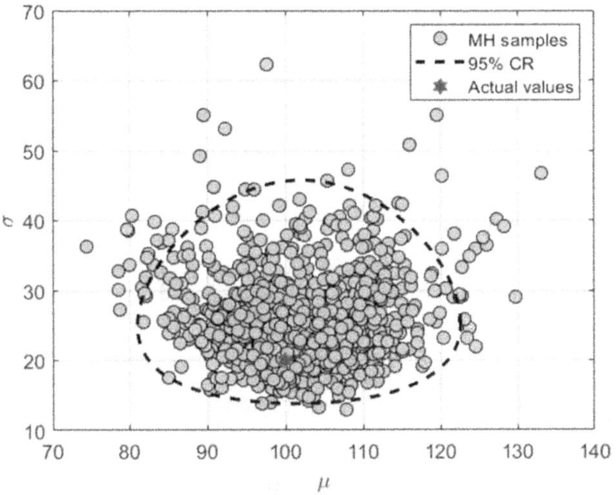

Figure 3.10 Posterior samples of μ and σ and the 95% confidence region (Example 3.7).

BAYESIAN PREDICTION

The X′ samples ~ f(x′|D) can be obtained using the similar procedure:

1. Randomly draw (μ,σ^2) from $\{(\mu_t,\sigma_t^2): t = T_b+1, \ldots, T\}$ and denote this sample by $\left(\hat{\mu},\hat{\sigma}^2\right)$.
2. Draw a sample $X' \sim N\left(\hat{\mu},\hat{\sigma}^2\right)$.
3. Cycle Steps 1 and 2 to obtain the desirable number of X′ sample.

3.3 BAYESIAN PROBLEMS WITH MULTIVARIATE NORMAL DATA

Although the content of the current section is applicable to Bayesian problems with multivariate normal data, for simplicity, only Bayesian problems with bivariate normal data are demonstrated. Now consider (X_1, X_2) data generated by a bivariate normal PDF with the following parameters:

$$\underline{\mu} = \begin{bmatrix} 100 \\ 50 \end{bmatrix} \qquad C = \begin{bmatrix} 20^2 & 0.8 \times 20 \times 5 \\ 0.8 \times 20 \times 5 & 5^2 \end{bmatrix} \tag{3.40}$$

Namely, $\mu_1 = 100$, $\sigma_1 = 20$, $\mu_2 = 50$, $\sigma_2 = 5$, and $\rho_{12} = 0.8$. Table 3.2 shows the generated data, and Figure 3.11 shows the data in the X_1–X_2 space. Given i.i.d. data $\mathbf{D} = \{\underline{X}^{(1)}, \underline{X}^{(2)}, \ldots, \underline{X}^{(m)}\}$ in Table 3.2, the purpose for parameter estimation is to estimate $\underline{\theta} = \{\mu_1, \sigma_1, \mu_2, \sigma_2, \rho_{12}\}$. This can be done by the maximum-likelihood method (frequentist) or by the Bayesian method.

Table 3.2 Sample values of (X_1, X_2) ($m = 10$). The underlying parameters are $\mu_1 = 100$, $\sigma_1 = 20$, $\mu_2 = 50$, $\sigma_2 = 5$, and $\delta_{12} = 0.8$

Simulated data		*Index*
$X_1^{(k)}$	$X_2^{(k)}$	k
124.06	60.88	1
113.35	54.18	2
134.75	50.96	3
90.52	48.92	4
133.54	57.72	5
74.40	45.29	6
81.15	41.40	7
105.41	48.06	8
71.43	42.74	9
89.81	41.70	10

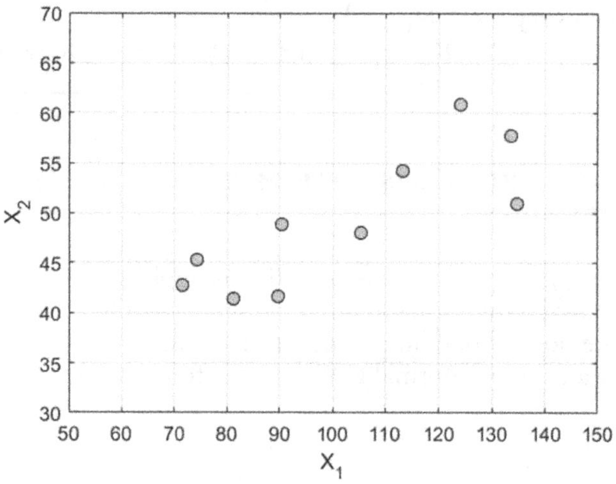

Figure 3.11 Sample values of (X_1, X_2) ($m = 10$).

EXAMPLE 3.8: Maximum-likelihood estimates with bivariate normal data

Assume $\underline{X} \sim N(\underline{\mu},\mathbf{C})$, where $\underline{\mu}$ and $\mathbf{C}$ are both unknown, where

$$\underline{\mu} = \begin{bmatrix} \mu_1 \\ \mu_2 \end{bmatrix} \quad \mathbf{C} = \begin{bmatrix} \sigma_1^2 & \rho_{12} \times \sigma_1 \times \sigma_2 \\ \rho_{12} \times \sigma_1 \times \sigma_2 & \sigma_2^2 \end{bmatrix} \tag{3.41}$$

Given i.i.d. data $\mathbf{D} = \{\underline{X}^{(1)}, \underline{X}^{(2)}, \ldots, \underline{X}^{(m)}\}$ in Table 3.2, the purpose is to estimate $\underline{\theta} = \{\mu_1, \sigma_1, \mu_2, \sigma_2, \rho_{12}\}$. The likelihood function can be expressed as

$$\begin{aligned} f(\mathbf{D} \mid \underline{\theta}) &= f\left(\underline{X}^{(1)}, \underline{X}^{(2)}, \ldots, \underline{X}^{(m)} \mid \mu_1, \mu_2, \sigma_1, \sigma_2, \rho_{12}\right) \\ &= \prod_{k=1}^{m} f\left(\underline{X}^{(k)} \mid \mu_1, \mu_2, \sigma_1, \sigma_2, \rho_{12}\right) = \prod_{k=1}^{m} \frac{e^{-0.5\times\left(\underline{X}^{(k)}-\underline{\mu}\right)^T C^{-1}\left(\underline{X}^{(k)}-\underline{\mu}\right)}}{\sqrt{2\pi}^2 \cdot \sqrt{|\mathbf{C}|}} \\ &= \frac{e^{-0.5\times\sum_{k=1}^{m}\left(\underline{X}^{(k)}-\underline{\mu}\right)^T C^{-1}\left(\underline{X}^{(k)}-\underline{\mu}\right)}}{(2\pi)^m \cdot |\mathbf{C}|^{m/2}} \end{aligned} \tag{3.42}$$

The log-likelihood function is

$$\begin{aligned} &\ln\left[f(\underline{X}^{(1)}, \underline{X}^{(2)}, \ldots, \underline{X}^{(m)} \mid \mu_1, \mu_2, \sigma_1, \sigma_2, \rho_{12})\right] \\ &= \sum_{k=1}^{m}\left[-\ln(2\pi) - 0.5 \times \ln\left(|\mathbf{C}|\right) - 0.5 \times \left(\underline{X}^{(k)} - \underline{\mu}\right)^T \mathbf{C}^{-1}\left(\underline{X}^{(k)} - \underline{\mu}\right)\right] \end{aligned} \tag{3.43}$$

The ML estimates for $\underline{\theta} = \{\mu_1, \sigma_1, \mu_2, \sigma_2, \rho_{12}\}$ can be obtained by maximizing the log-likelihood with the constraints $\sigma_1 > 0$, $\sigma_2 > 0$, and $-1 \le \rho_{12} \le 1$. This requires a constrained optimization method (such as the fmincon.m function in MATLAB). The resulting ML estimates are $\mu_{1,ML} = 101.84$, $\sigma_{1,ML} = 22.60$, $\mu_{2,ML} = 49.18$, $\sigma_{2,ML} = 6.42$, and $\rho_{12,ML} = 0.83$.

EXAMPLE 3.9: Bayesian analysis with unknown $\underline{\mu}$ but known C (parameter estimation only)

Assume $\underline{X} \sim N(\underline{\mu}, \mathbf{C})$, where $\underline{\mu}$ is unknown, but $\mathbf{C}$ is known. The prior PDF of $\underline{\mu}$ is assumed to follow the bivariate normal PDF: $\underline{\mu} \sim N(\underline{\mu}_0, \mathbf{C}_0)$, where $\underline{\mu}_0$ and $\mathbf{C}_0$ are specified based on prior information on $\underline{\mu}$. Given i.i.d. data $\mathbf{D} = \{\underline{X}^{(1)}, \underline{X}^{(2)}, \ldots, \underline{X}^{(m)}\}$

in Table 3.2 and the knowledge of the actual **C**, the purpose is to calculate the posterior PDF of $\underline{\mu}$. The prior PDF of $\underline{\mu}$ has the following expression:

$$f(\underline{\mu}) = \frac{1}{\sqrt{2\pi}^{2} \cdot |\mathbf{C}_0|} \exp\left[-0.5 \times (\underline{\mu} - \underline{\mu}_0)^T \mathbf{C}_0^{-1} (\underline{\mu} - \underline{\mu}_0)\right] \quad (3.44)$$

The likelihood function of $\underline{\mu}$ has the following expression:

$$f(\mathbf{D} \mid \underline{\mu}) = \frac{1}{(2\pi)^m \cdot |\mathbf{C}|^{m/2}} \exp\left[-0.5 \times \sum_{k=1}^{m} (\underline{X}^{(k)} - \underline{\mu})^T \mathbf{C}^{-1} (\underline{X}^{(k)} - \underline{\mu})\right] \quad (3.45)$$

The posterior PDF of $\underline{\mu}$ is proportional to the product of the likelihood function of $\underline{\mu}$ and the prior PDF of $\underline{\mu}$. Note that the multivariate normal prior PDF of $\underline{\mu}$ is conjugate to the multivariate normal likelihood function of $\underline{\mu}$ (see Section 2.5). According to Table 2.1, the posterior PDF of $\underline{\mu}$ is still multivariate normal:

$$f(\underline{\mu} \mid \mathbf{D}) = N\left[(\mathbf{C}_0^{-1} + m\mathbf{C}^{-1})^{-1} \left(\mathbf{C}_0^{-1}\underline{\mu}_0 + \mathbf{C}^{-1}\sum_{k=1}^{m} \underline{X}^{(k)}\right), (\mathbf{C}_0^{-1} + m\mathbf{C}^{-1})^{-1}\right] = N(\underline{\mu}'_0, \mathbf{C}'_0) \quad (3.46)$$

where

$$\underline{\mu}'_0 = (\mathbf{C}_0^{-1} + m\mathbf{C}^{-1})^{-1} \left(\mathbf{C}_0^{-1}\underline{\mu}_0 + \mathbf{C}^{-1}\sum_{k=1}^{m} \underline{X}^{(k)}\right) \quad \mathbf{C}'_0 = (\mathbf{C}_0^{-1} + m\mathbf{C}^{-1})^{-1} \quad (3.47)$$

EXAMPLE 3.10: Bayesian analysis with known $\underline{\mu}$ but unknown C (parameter estimation only)

Assume $\underline{X} \sim N(\underline{\mu}, \mathbf{C})$, where $\underline{\mu}$ is known, but **C** is unknown. The prior PDF of **C** is assumed to follow the inverse-Wishart PDF: $\mathbf{C} \sim IW(\Sigma_0, \nu_0)$, where Σ_0 and ν_0 are specified based on prior information on **C**. Given i.i.d. data $\mathbf{D} = \{\underline{X}^{(1)}, \underline{X}^{(2)}, \ldots, \underline{X}^{(m)}\}$ in Table 3.2 and the knowledge of the actual $\underline{\mu}$, the purpose is to calculate the posterior PDF of **C**. The prior PDF of **C** has the following expression:

$$f(\mathbf{C}) = \frac{|\Sigma_0|^{\nu_0/2}}{2^{\nu_0 n/2}\Gamma_n(\nu_0/2)} |\mathbf{C}|^{-(\nu_0+n+1)/2} \exp\left[-0.5 \times tr(\Sigma_0 \times \mathbf{C}^{-1})\right] \quad (3.48)$$

The likelihood function of **C** has the following expression:

$$f(\mathbf{D} \mid \mathbf{C}) = \frac{1}{\sqrt{2\pi}^{mn} \cdot \left|\mathbf{C}\right|^{m/2}} \exp\left[-0.5 \times \sum_{k=1}^{m} \left(\underline{X}^{(k)} - \underline{\mu}\right)^{T} \mathbf{C}^{-1} \left(\underline{X}^{(k)} - \underline{\mu}\right)\right] \quad (3.49)$$

The posterior PDF of **C** is proportional to the product of the likelihood function of **C** and the prior PDF of **C**. Note that the prior inverse-Wishart PDF of **C** is conjugate to the multivariate normal likelihood function of **C** (see Section 2.5). According to Table 2.1, the posterior PDF of **C** is still inverse-Wishart:

$$f(\mathbf{C} \mid \mathbf{D}) = IW\left(\Sigma_0 + \sum_{k=1}^{m} \left(\underline{X}^{(k)} - \underline{\mu}\right)\left(\underline{X}^{(k)} - \underline{\mu}\right)^{T}, m + \nu_0\right) = IW\left(\Sigma'_0, \nu'_0\right) \quad (3.50)$$

where

$$\Sigma'_0 = \Sigma_0 + \sum_{k=1}^{m} \left(\underline{X}^{(k)} - \underline{\mu}\right)\left(\underline{X}^{(k)} - \underline{\mu}\right)^{T} \qquad \nu'_0 = m + \nu_0 \quad (3.51)$$

EXAMPLE 3.11: Bayesian analysis with unknown μ and C (parameter estimation and prediction with the GS algorithm)

Assume $\underline{X} \sim N(\underline{\mu}, \mathbf{C})$, where $\underline{\mu}$ and **C** are both unknown. The prior PDF of $\underline{\mu} \sim N(\underline{\mu}_0, \mathbf{C}_0)$, and the prior PDF of $\mathbf{C} \sim IW(\Sigma_0, \nu_0)$. For (roughly) non-informative priors, the following prior parameters are adopted: $\underline{\mu}_0 = (0\ 0)^T$, $\mathbf{C}_0 = 10{,}000^2 \times \mathbf{I}_{2\times2}$, $\Sigma_0 = \mathbf{I}_{2\times2}$, and $\nu_0 = n+1 = 3$. Note that $IW(\Sigma_0, \nu_0)$ with $\Sigma_0 = \mathbf{I}_{2\times2}$ and $\nu_0 = n+1$ is only weakly informative (see Section 2.4.3). A more sophisticated way of making the IW PDF more non-informative will be discussed in Chapter 4.

BAYESIAN PARAMETER ESTIMATION

Given i.i.d. data $\mathbf{D} = \{\underline{X}^{(1)}, \underline{X}^{(2)}, \ldots, \underline{X}^{(m)}\}$ in Table 3.2, the purpose is to draw $(\underline{\mu}, \mathbf{C})$ samples $\sim f(\underline{\mu}, \mathbf{C} \mid \mathbf{D})$. Due to the conjugate priors, the fully conditional PDFs have analytical expressions (see Table 2.1):

$$f(\underline{\mu} \mid \mathbf{C}, \mathbf{D}) = N\left(\left(\mathbf{C}_0^{-1} + m\mathbf{C}^{-1}\right)^{-1}\left(\mathbf{C}_0^{-1}\underline{\mu}_0 + \mathbf{C}^{-1}\sum_{k=1}^{m}\underline{X}^{(k)}\right), \left(\mathbf{C}_0^{-1} + m\mathbf{C}^{-1}\right)^{-1}\right) \quad (3.52)$$

$$f(\mathbf{C}\,|\,\underline{\mu},\mathbf{D}) = \mathrm{IW}\left(\Sigma_0 + \sum_{k=1}^{m}\left(\underline{X}^{(k)} - \underline{\mu}\right)\left(\underline{X}^{(k)} - \underline{\mu}\right)^T, m + \nu_0\right) \tag{3.53}$$

Therefore, the following Gibbs sampler (GS) algorithm can be adopted:

1. Initialize $(\underline{\mu}_{t=0},\mathbf{C}_{t=0})$ at any values.
2. Update sample $\underline{\mu}_{t=1} \sim N\left(\left(\mathbf{C}_0^{-1} + m\mathbf{C}_{t=0}^{-1}\right)^{-1}\left(\mathbf{C}_0^{-1}\underline{\mu}_0 + \mathbf{C}_{t=0}^{-1}\sum_{k=1}^{n}\underline{X}^{(k)}\right), \left(\mathbf{C}_0^{-1} + m\mathbf{C}_{t=0}^{-1}\right)^{-1}\right)$
3. Update sample $\mathbf{C}_{t=1} \sim \mathrm{IW}\left(\Sigma_0 + \sum_{k=1}^{m}\left(\underline{X}^{(k)} - \underline{\mu}_{t=1}\right)\left(\underline{X}^{(k)} - \underline{\mu}_{t=1}\right)^T, m + \nu_0\right)$
4. Cycle Steps 2 and 3 to obtain MC samples $\{(\underline{\mu}_t,\mathbf{C}_t): t = 0, \ldots, T\}$.

With $T = 11{,}000$ and $T_b = 1000$, Figure 3.12 shows the posterior samples of $(\underline{\mu},\mathbf{C})$ in terms of $(\mu_1, \sigma_1, \mu_2, \sigma_2, \rho_{12})$. The 95% CRs are also plotted by assuming that (μ_1, μ_2) samples follow the bivariate normal PDF and that $[\ln(\sigma_1), \ln(\sigma_2)]$ samples follow the bivariate normal PDF.

BAYESIAN PREDICTION

The $\underline{X}'$ samples $\sim f(\underline{x}'|\mathbf{D})$ can be obtained using the following procedure:

1. Randomly draw $(\underline{\mu},\mathbf{C})$ from $\{(\underline{\mu}_t,\mathbf{C}_t): t = T_b+1, \ldots, T\}$ and denote this sample by $\left(\hat{\underline{\mu}},\hat{\mathbf{C}}\right)$.
2. Draw a sample $\underline{X}' \sim N\left(\hat{\underline{\mu}},\hat{\mathbf{C}}\right)$.
3. Cycle Steps 1 and 2 to obtain the desirable number of $\underline{X}'$ sample.

Figure 3.13 shows the resulting $\underline{X}'$ samples $\sim f(\underline{x}'|\mathbf{D})$. The 95% CR is plotted by assuming that (X'_1, X'_2) samples follow the bivariate normal PDF. The original data $\mathbf{D} = \{\underline{X}^{(1)}, \underline{X}^{(2)}, \ldots, \underline{X}^{(m)}\}$ in Table 3.2 are shown as grey circles in the figure for comparison.

If more $\underline{X}$ data are generated ($m = 100$ data points), the likelihood function becomes more peaked, so the posterior samples become more concentrative. Figure 3.14 shows the posterior samples of $(\underline{\mu},\mathbf{C})$ in terms of $(\mu_1, \sigma_1, \mu_2, \sigma_2, \rho_{12})$, whereas Figure 3.15 shows the resulting $\underline{X}'$ samples $\sim f(\underline{x}'|\mathbf{D})$.

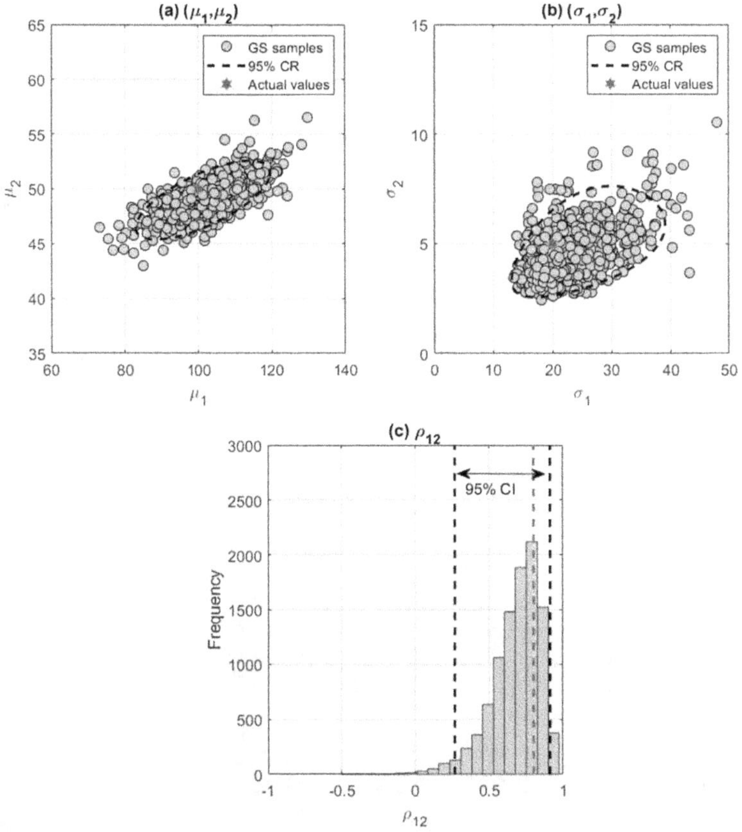

Figure 3.12 Posterior samples of $\underline{\mu}$ and **C** and the 95% confidence region (Example 3.11, m = 10).

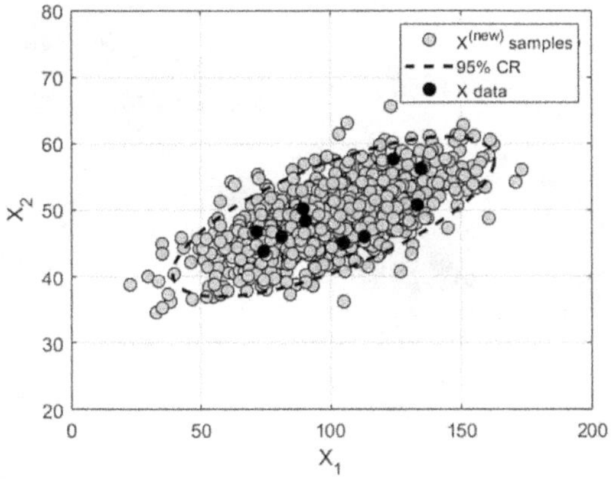

Figure 3.13 Posterior $\underline{X}'$ samples ~ f($\underline{x}'$|**D**) (Example 3.11, m = 10).

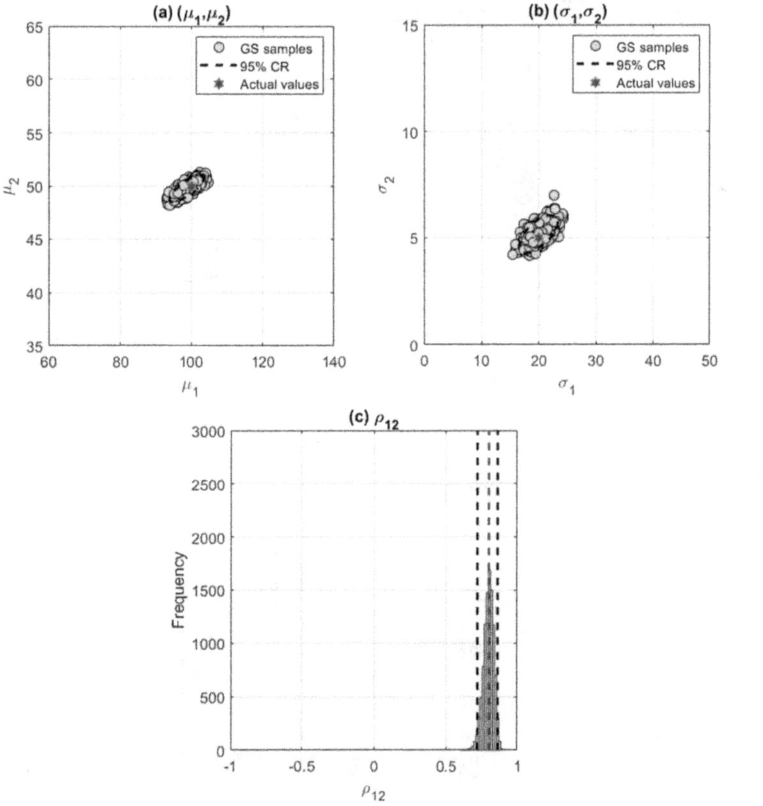

Figure 3.14 Posterior samples of $\underline{\mu}$ and **C** and the 95% confidence region (Example 3.11, m = 100).

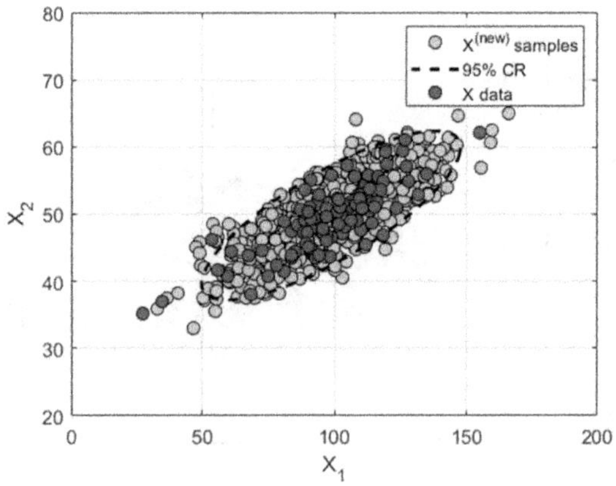

Figure 3.15 Posterior $\underline{X}'$ samples ~ f($\underline{x}'$|**D**) (Example 3.11, m = 100).

3.4 BAYESIAN PROBLEMS WITH SPATIAL VARIABILITY

Now consider a set of spatial data of soil property $\xi(z) = t(z) + w(z)$ simulated at the depth range of 0–20 m with 0.1 m sampling interval $z_1 = 0.1$ m, $z_2 = 0.2$ m, …, $z_{200} = 20$ m (there are m = 200 depths). Suppose that the underlying trend is a linear trend:

$$t(z) = 100 + 5 \cdot z \tag{3.54}$$

The spatial variability w(z) is a zero-mean stationary normal random field with standard deviation $\sigma = 10$. The auto-correlation function is single exponential with scale of fluctuation (SOF) $\delta = 0.5$ m:

$$\rho(\Delta z) = \exp\left(-2 \cdot |\Delta z| / \delta\right) \tag{3.55}$$

Figure 3.16 shows the simulated data with respect to depth, $\xi(z_1)$, $\xi(z_2)$, … $\xi(z_m)$. Given data $\underline{D} = \{\xi(z_1)\ \xi(z_2) \ldots \xi(z_m)\}$, the purpose of parameter estimation is to estimate the auto-covariance parameters (such as variance and auto-correlation parameters) for the spatial variability, denoted by $\underline{\theta}^{(w)}$, and the

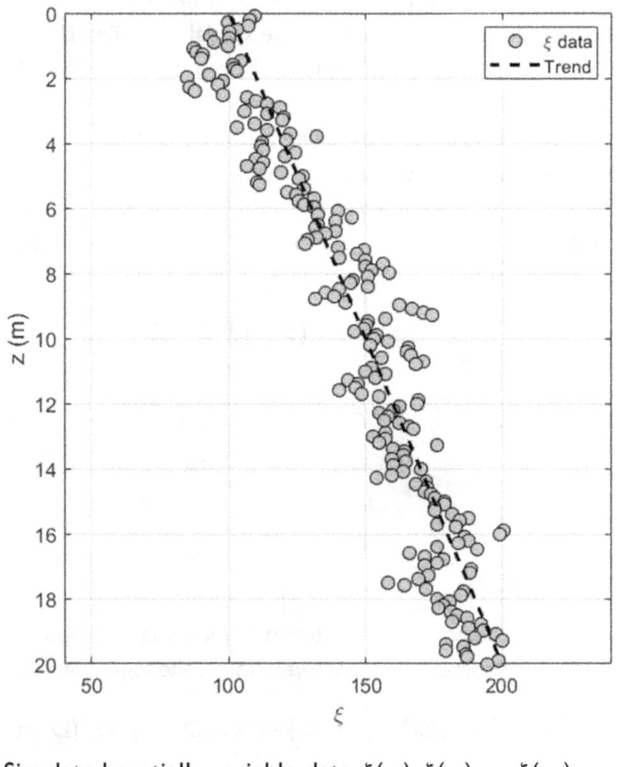

Figure 3.16 Simulated spatially variable data, $\xi(z_1)$, $\xi(z_2)$, … $\xi(z_m)$.

trend parameters {a,b}, denoted by $\underline{\theta}^{(t)}$, where the superscripts "(w)" and "(t)" denote the spatial variability and trend, respectively. Let $\underline{\theta} = \{\underline{\theta}^{(w)},\underline{\theta}^{(t)}\}$ denote the collection of all unknown parameters. The estimation of $\underline{\theta}$ can be done by the maximum-likelihood method (frequentist) or by the Bayesian method.

EXAMPLE 3.12: Maximum-likelihood estimates for spatial variability data

Let us consider the linear trend model:

$$t(z) = a + b \cdot z \tag{3.56}$$

but the trend parameters a and b are unknown. Let us also model the spatial variability w(z) by a zero-mean stationary normal random field with the Whittle–Matérn (WM) model (see Section 2.6) as the auto-correlation model:

$$\rho(\Delta z) = \frac{2}{\Gamma(\nu)} \cdot \left(\frac{\sqrt{\pi} \cdot \Gamma(\nu + 0.5) \cdot |\Delta z|}{\Gamma(\nu) \cdot \delta} \right)^{\nu} K_{\nu} \left(\frac{2\sqrt{\pi} \cdot \Gamma(\nu + 0.5) \cdot |\Delta z|}{\Gamma(\nu) \cdot \delta} \right) \tag{3.57}$$

but the auto-correlation parameters δ (scale of fluctuation) and ν (smoothness) are unknown. The variance (σ^2) of the spatial variability is also unknown. Given data $\underline{D} = \{\xi(z_1)\ \xi(z_2)\ \ldots\ \xi(z_m)\}$, the purpose is to find the ML estimates of $\underline{\theta} = \{\underline{\theta}^{(w)},\underline{\theta}^{(t)}\} = \{\sigma, \delta, \nu, a, b\}$.

Note that $w(z) = \xi(z) - t(z)$ is modeled as the zero-mean stationary normal random field. This means that $\underline{D} = [\xi(z_1)\ \xi(z_2)\ \ldots\ \xi(z_m)]^T$ follow a multivariate normal PDF with mean vector = $\underline{t} = [t(z_1)\ t(z_2)\ \ldots\ t(z_m)]^T$ and autocovariance matrix same as that for $\underline{w} = [w(z_1)\ w(z_2)\ \ldots\ w(z_m)]^T$. As a result, the likelihood function has the following expression:

$$f(\underline{D} \mid \underline{\theta}) = \frac{1}{\sqrt{2\pi}^{m} \cdot \sqrt{|\mathbf{C}|}} \cdot \exp\left[-0.5 \times (\underline{D} - \underline{t})^T \mathbf{C}^{-1} (\underline{D} - \underline{t}) \right] \tag{3.58}$$

where

$$\mathbf{C} = \sigma^2 \times \begin{bmatrix} 1 & \rho(z_1 - z_2) & \cdots & \rho(z_1 - z_m) \\ & 1 & \cdots & \rho(z_2 - z_m) \\ & & & \vdots \\ \text{symmetric} & & & 1 \end{bmatrix} \tag{3.59}$$

where $\rho(.)$ is the auto-correlation for the WM model in Equation (3.57), which depends on the auto-correlation parameters (δ, ν). The log-likelihood function is

$$\ln[f(\underline{D} \mid \underline{\theta})] = -\frac{m}{2} \times \ln(2\pi) - \frac{1}{2} \ln(|\mathbf{C}|) - 0.5 \times (\underline{D} - \underline{t})^T \mathbf{C}^{-1} (\underline{D} - \underline{t}) \tag{3.60}$$

The ML estimates for $\underline{\theta} = \{\sigma, \delta, \nu, a, b\}$ can be obtained by maximizing the log-likelihood with the constraints $\sigma > 0$, $\delta > 0$, and $\nu > 0$. Caution must be taken when evaluating the term $\ln(|\mathbf{C}|)$. The determinant $|\mathbf{C}|$ can easily approach zero when the number of data points is large. This can be mitigated by first determining the Cholesky decomposition for $\mathbf{C}$: $\mathbf{C} = \mathbf{L}\times\mathbf{L}^T$, where $\mathbf{L}$ is a lower triangular Cholesky decomposition. It follows that

$$\ln[f(\underline{D} \mid \underline{\theta})] = -\frac{m}{2} \times \ln(2\pi) - \sum_{i=1}^{m} \ln(L_{ii}) - \frac{1}{2} \underline{r}^T \underline{r} \tag{3.61}$$

where L_{ii} is the ith diagonal in $\mathbf{L}$; $\underline{r}$ is the solution of the linear equations $\mathbf{L}\times\underline{r} = \underline{D} - \underline{t}$, and $\underline{r}$ can be solved efficiently using forward substitution because $\mathbf{L}$ is a lower triangular matrix. The resulting ML estimates are $a_{ML} = 98.7$, $b_{ML} = 4.90$, $\sigma_{ML} = 10.0$, $\delta_{ML} = 0.59$, and $\nu_{ML} = 0.43$. The actual underlying parameters $a = 100$, $b = 5$, $\sigma = 10$, $\delta = 0.5$, and $\nu = 0.5$ (note: the underlying single exponential model has $\nu = 0.5$).

The Bayesian analysis for the spatial data is challenging. First, there are no analytical Bayesian solutions for the posterior PDF $f(\underline{\theta}|\underline{D})$. Second, there are no conjugate priors, so the GS algorithm is inconvenient. Finally, there are five unknown parameters, so for the MH algorithm, there are five proposal PDF standard deviations ($\sigma_{p,a}$, $\sigma_{p,b}$, …, $\sigma_{p,\nu}$) to tune. It can be fairly challenging to fine-tune the MH algorithm such that it has a desirable Markov chain behavior. In the case where both NH and GS algorithms are not appropriate, a more advanced variant of MCMC methods, called the transitional Markov chain Monte Carlo (TMCMC) method (Ching and Chen 2007), may be adopted. Note that the TMCMC algorithm is also applicable to general Bayesian problems not related to spatial variability although it is presented in the current section to solve Bayesian problems with spatial variability.

3.4.1 Transitional Markov chain Monte Carlo algorithm

The TMCMC algorithm (Ching and Chen 2007) is a variant of MCMC methods. It does not require conjugate priors (as for the GS algorithm). It also does not require the specification of proposal PDFs (as for the MH algorithm) because the proposal PDF is constructed automatically during the algorithm. The basic idea of the TMCMC algorithm is to sample from a series of intermediate PDFs that converge to the posterior $f(\underline{\theta}|\underline{D})$ from the prior $f(\underline{\theta})$:

$$f_j(\underline{\theta}) \propto f(\underline{D} \mid \underline{\theta})^{p_j} f(\underline{\theta}) \quad j = 0,\ldots,M \quad 0 = p_0 < p_1 < \ldots < p_M = 1 \tag{3.62}$$

where the index j denotes the stage number. It is clear that $f_0(\underline{\theta})$ is the same as the prior PDF $f(\underline{\theta})$, whereas $f_M(\underline{\theta})$ is the same as the posterior PDF $f(\underline{\theta}|\underline{D})$. The TMCMC algorithm starts from j = 0 ($p_0 = 0$) by drawing N $\underline{\theta}$ samples from the prior PDF $f(\underline{\theta}) = f_0(\underline{\theta})$, denoted by $\{\underline{\theta}_{0,k}: k = 1, ..., N\}$. The kth sample $\underline{\theta}_{0,k}$ is attached to the following sampling-importance weight:

$$w_{0,k} = \frac{f_1(\underline{\theta}_{0,k})}{f_0(\underline{\theta}_{0,k})} = \frac{f(\underline{D} \mid \underline{\theta}_{0,k})^{p_1} f(\underline{\theta}_{0,k})}{f(\underline{D} \mid \underline{\theta}_{0,k})^{p_0} f(\underline{\theta}_{0,k})} = f(\underline{D} \mid \underline{\theta}_{0,k})^{p_1 - p_0} \tag{3.63}$$

The $\{\underline{\theta}_{0,k}: k = 1, ..., N\}$ samples are then resampled according to their weights $\{w_{0,k}: k = 1, ..., N\}$ to yield the $\underline{\theta}$ samples ~ $f_1(\underline{\theta})$, which is called the sampling-importance resampling (SIR) procedure (Rubin 1988). To avoid repetitive $\underline{\theta}$ samples after the resampling, a simple solution is to conduct the MH algorithm on each resampled sample with the stationary PDF equal to $f_1(\underline{\theta})$. More precisely, with probability $w_{0,k}/(\Sigma w_{0,k})$, a Markov chain sample in the kth chain (the chain with $\underline{\theta}_{0,k}$ as the leader) is generated using a multivariate normal proposal PDF centered at the current sample of the kth chain with a covariance matrix equal to the scaled version of the estimated covariance matrix of $f_1(\underline{\theta})$:

$$\Sigma = \beta^2 \times \left\{ \sum_{k=1}^{N} w_{0,k} \left(\underline{\theta}_{0,k} - \left[\sum_{m=1}^{N} w_{0,m} \underline{\theta}_{0,m} \middle/ \left(\sum_{n=1}^{N} w_{0,n} \right) \right] \right) \right. \\ \left. \left(\underline{\theta}_{0,k} - \left[\sum_{m=1}^{N} w_{0,m} \underline{\theta}_{0,m} \middle/ \left(\sum_{n=1}^{N} w_{0,n} \right) \right] \right)^T \right\} \middle/ \left(\sum_{k=1}^{N} w_{0,k} \right) \tag{3.64}$$

where β is a prescribed scaling factor. The β value should be chosen to suppress the rejection rate of the MH step and, at the same time, to make large MH moves. It is found that 0.2~0.5 is a reasonable choice of the scaling parameter β. The SIR is done N times to obtain $\{\underline{\theta}_{1,k}: k = 1, ..., N\}$. Note that we always generate new Markov chain samples at the end of each chain. For example, if the kth chain is chosen for the third time, we generate the new sample using the second sample in that chain as the current sample, rather than using the leader $\underline{\theta}_{0,k}$. The $\{\underline{\theta}_{1,k}: k = 1, ..., N\}$ samples TMCMC algorithm are then resampled according to their weights $\{w_{1,k}: k = 1, ..., N\}$ to yield the $\underline{\theta}$ samples ~ $f_2(\underline{\theta})$, where the weight $w_{1,k}$ is computed using a formula similar to Equation (3.63), but the power of the likelihood becomes (p_2-p_1). This process repeats for stage j = 0, 1, ..., M. At the Mth stage, $p_M = 1$, so the resulting $\{\underline{\theta}_{M,k}: k = 1, ..., N\}$ samples are distributed as $f_M(\underline{\theta}) \propto f(\underline{D}|\underline{\theta})f(\underline{\theta})$, which is the posterior PDF $f(\underline{\theta}|\underline{D})$.

The choice of $\{p_1, p_2, ..., p_{M-1}\}$ is essential. It is desirable to increase the p values slowly so that the transition between adjacent PDFs is smooth, but if the increase of the p values is too slow, the required number of intermediate stages (i.e., M value) will be large, which is also not desirable. Notice that the degree of uniformity the $\{w_{0,k}: k = 1, ..., N\}$ is a good indicator of how close $f_1(\underline{\theta})$ is to $f_0(\underline{\theta})$, so p_1 should be chosen so that the coefficient of variation (COV) of the weights is equal to a prescribed threshold. Ching and Chen (2007) found that 100% is usually a reasonable choice of this threshold.

EXAMPLE 3.13: Bayesian analysis for spatial variability data

Let us consider the same trend model as in Example 3.12. Let us also adopt the WM model as the auto-correlation model. Let us adopt the following prior PDFs for $\underline{\theta} = \{\underline{\theta}^{(t)},\underline{\theta}^{(w)}\} = \{a, b, \sigma, \delta, \nu\}$: $a \sim$ unif[0, 500], $b \sim$ unif[−20, 20], $\ln(\sigma) \sim$ unif[ln(0.1), ln(100)], $\ln(\delta) \sim$ unif[ln(0.01), ln(10)], and $\ln(\nu) \sim$ unif[ln(0.1), ln(3)]. Other types of prior PDFs can also be used. The uniform priors are adopted for demonstration only. Also note that logarithms are taken for σ, δ, and ν for convenience because these parameters cannot be negative. The sample size for each TMCMC stage is taken to be N = 2000.

BAYESIAN PARAMETER ESTIMATION

Given data $\underline{D} = \{\xi(z_1), \xi(z_2) ..., \xi(z_m)\}$, the purpose is to draw $\underline{\theta}$ samples $\sim f(\underline{\theta}|\underline{D})$. The resulting TMCMC samples $\{(a_k,b_k,\sigma_k,\delta_k,\nu_k): k = 1, ..., N\}$ are shown in Figure 3.17.

BAYESIAN PREDICTION

Now consider $\underline{D}' = \{\xi'(z_1), \xi'(z_2), ..., \xi'(z_m)\}$ that follows the same trend, standard deviation, SOF, and smoothness as the original data $\underline{D} = \{\xi(z_1), \xi(z_2), ..., \xi(z_m)\}$. Namely, $\xi(z) = t(z) + w(z)$ (original data) and $\xi'(z) = t'(z) + w'(z)$ (new data), where the trends $t(z)$ and $t'(z)$ are the same, yet $w(z)$ and $w'(z)$ are independent and identically distributed (w and w′ follow the same standard deviation, SOF, and smoothness). The purpose of the Bayesian prediction is to draw $\underline{D}'$ samples $\sim f(\underline{D}'|\underline{D})$. According to the Total Probability Theorem and the Law of Large Numbers:

$$f(\underline{D}' \mid \underline{D}) = \int f(\underline{D}' \mid \underline{\theta}, \underline{D}) f(\underline{\theta} \mid \underline{D}) d\underline{\theta} = \int f(\underline{D}' \mid \underline{\theta}) f(\underline{\theta} \mid \underline{D}) d\underline{\theta} \approx \frac{1}{N} \sum_{k=1}^{N} f(\underline{D}' \mid \underline{\theta}_k) \quad (3.65)$$

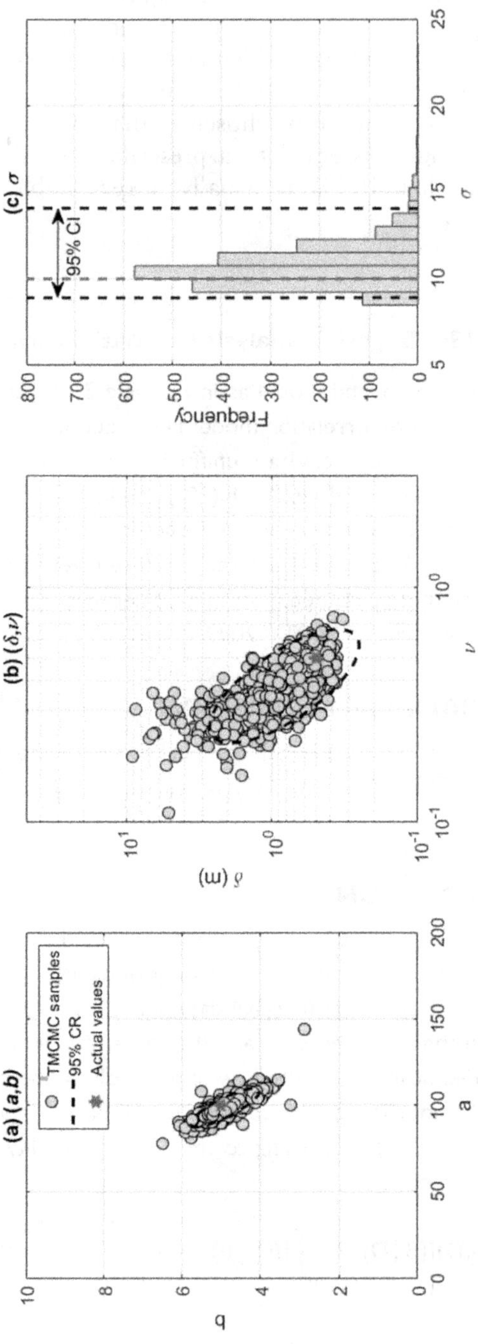

Figure 3.17 Posterior samples of $\underline{\theta}$ and the 95% confidence region (Example 3.13).

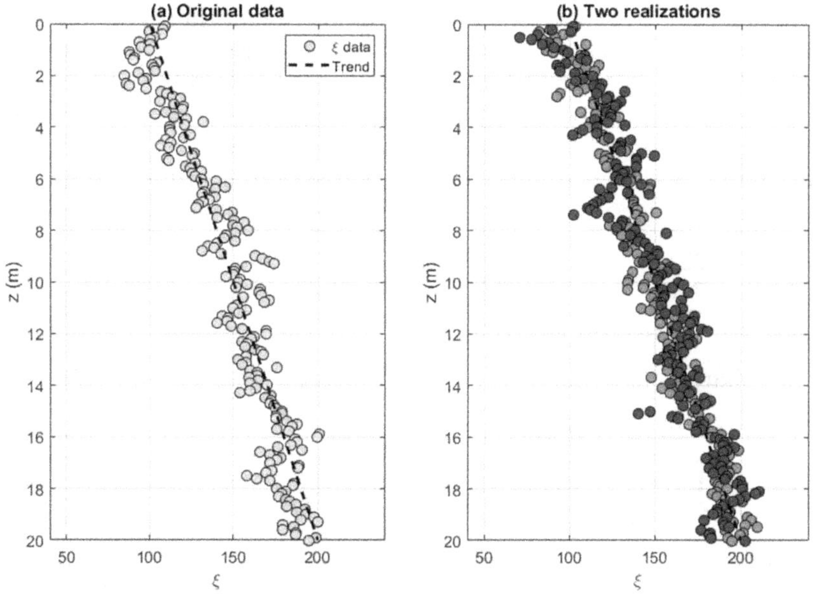

Figure 3.18 $\underline{D}'$ samples ~ $f(\underline{D}'|\underline{D})$ (Example 3.13).

where $\underline{\theta}_k = \{a_k, b_k, \sigma_k, \delta_k, \nu_k\}$ is the kth sample drawn from $f(\underline{\theta}|\underline{D})$, that is, the kth TMCMC samples at the final stage. As a result, $\underline{D}'$ samples ~ $f(\underline{D}'|\underline{D})$ can be obtained using the following procedure:

1. Randomly draw $\underline{\theta} = \{a,b,\sigma,\delta,\nu\}$ from the TMCMC samples $\{\underline{\theta}_{M,k}: k = 1, \ldots, N\}$ ("M" denotes the final stage of TMCMC) and denote these samples by $\left(\hat{a}, \hat{b}, \hat{\sigma}, \hat{\delta}, \hat{\nu}\right)$.
2. Based on $\left(\hat{\sigma}, \hat{\delta}, \hat{\nu}\right)$, one can obtain the autocovariance matrix $\hat{\mathbf{C}} = \hat{\sigma}^2 \times \hat{\mathbf{R}}$. Note that the auto-correlation matrix $\hat{\mathbf{R}}$ can be computed based on the auto-correlation parameters $\left(\hat{\delta}, \hat{\nu}\right)$. Draw a sample $\underline{w}' \sim N\left(\underline{0}, \hat{\mathbf{C}}\right)$.
3. Based on $\left(\hat{a}, \hat{b}\right)$, one can obtain the trend vector $\underline{t}'$. A sample of $\underline{D}'$ can be obtained by $\underline{D}' = \underline{t}' + \underline{w}'$.
4. Cycle Steps 1–3 to obtain the desirable number of $\underline{D}'$ samples.

Figure 3.18b shows two $\underline{D}'$ realizations, and Figure 3.18a shows the original $\underline{D}$ for comparison.

3.4.2 Bayesian model selection

A useful by-product produced by the TMCMC algorithm is the (Bayesian) evidence $f(\underline{D})$. The Bayesian evidence is useful for the purpose of Bayesian model selection (Beck and Yuen 2004). To see how the TMCMC algorithm can estimate the evidence $f(\underline{D})$ as a by-product, let us denote the sample mean of the weights $\{w_{j,k}: k = 1, \ldots, N\}$ by S_j. Note that $w_{j,k} = f(\underline{D}|\underline{\theta}_{j,k})^{pj+1-pj}$ is computed based on samples $\underline{\theta}_{j,k} \sim f_j(\underline{\theta})$. According to the Law of Large Numbers, S_j is an estimate of the following expectation:

$$S_j = \frac{1}{N}\sum_{k=1}^{N} w_{j,k} \approx E\left(w_{j,.}\right) = \int f(\underline{D} \mid \underline{\theta})^{P_{j+1}-P_j} f_j\left(\underline{\theta}\right)d\underline{\theta} \tag{3.66}$$

Note that

$$f_j\left(\underline{\theta}\right) = \frac{f(\underline{D} \mid \underline{\theta})^{P_j} f\left(\underline{\theta}\right)}{\int f(\underline{D} \mid \underline{\theta})^{P_j} f\left(\underline{\theta}\right)d\underline{\theta}} \tag{3.67}$$

As a result,

$$S_j \approx \int f(\underline{D} \mid \underline{\theta})^{P_{j+1}-P_j} \frac{f(\underline{D} \mid \underline{\theta})^{P_j} f\left(\underline{\theta}\right)}{\int f(\underline{D} \mid \underline{\theta})^{P_j} f\left(\underline{\theta}\right)d\underline{\theta}} d\underline{\theta} = \frac{\int f(\underline{D} \mid \underline{\theta})^{P_{j+1}} f\left(\underline{\theta}\right)d\underline{\theta}}{\int f(\underline{D} \mid \underline{\theta})^{P_j} f\left(\underline{\theta}\right)d\underline{\theta}} \tag{3.68}$$

It is then clear that the evidence $f(\underline{D})$ can be estimated as the series product $S_0 \times S_1 \times \ldots \times S_{M-1}$:

$$\begin{aligned} f\left(\underline{D}\right) &= \int f\left(\underline{D} \mid \underline{\theta}\right) f\left(\underline{\theta}\right)d\underline{\theta} = \int f\left(\underline{D} \mid \underline{\theta}\right)^{PM} f\left(\underline{\theta}\right)d\underline{\theta} \\ &= \frac{\int f\left(\underline{D} \mid \underline{\theta}\right)^{P_1} f\left(\underline{\theta}\right)d\underline{\theta}}{\int f\left(\underline{D} \mid \underline{\theta}\right)^{P_0} f\left(\underline{\theta}\right)d\underline{\theta}} \times \frac{\int f\left(\underline{D} \mid \underline{\theta}\right)^{P_1} f\left(\underline{\theta}\right)d\underline{\theta}}{\int f\left(\underline{D} \mid \underline{\theta}\right)^{P_2} f\left(\underline{\theta}\right)d\underline{\theta}} \times \cdots \\ &\times \frac{\int f\left(\underline{D} \mid \underline{\theta}\right)^{PM} f\left(\underline{\theta}\right)d\underline{\theta}}{\int f\left(\underline{D} \mid \underline{\theta}\right)^{PM-1} f\left(\underline{\theta}\right)d\underline{\theta}} \approx S_0 \times S_1 \times \cdots \times S_{M-1} \end{aligned} \tag{3.69}$$

The Bayesian model selection can be conducted for any Bayesian problem with competing models. For instance, for a given i.i.d. data $\underline{D} = \{X^{(1)}, X^{(2)}, \ldots, X^{(m)}\}$, one can analyze the data by assuming the underlying distribution is the normal PDF or by assuming the underlying distribution is the lognormal PDF. One can execute the TMCMC algorithm twice (one for the normal PDF model and the other for the lognormal PDF model) to estimate the evidence of the two competing models, denoted by $f(\underline{D}|\mathbf{M}_N)$ and $f(\underline{D}|\mathbf{M}_{LN})$, where the subscripts "N" and "LN" denote the normal and lognormal PDF models, respectively. The model that gives larger evidence $f(\underline{D}|\mathbf{M})$ is a more plausible model. Suppose that there are several competing models $\mathbf{M}_1$, $\mathbf{M}_2$, …, $\mathbf{M}_p$ for the data $\underline{D}$. The TMCMC algorithm is executed for each model $\mathbf{M}_i$, and the evidence $f(\underline{D}|\mathbf{M}_i)$ can be estimated for each model $\mathbf{M}_i$. One can then estimate the posterior probability for each model, denoted by $P(\mathbf{M}_i|\underline{D})$, based on the Bayes Theorem:

$$P(\mathbf{M}_i \mid \underline{D}) = \frac{f(\underline{D} \mid \mathbf{M}_i)P(\mathbf{M}_i)}{\sum_{j=1}^{p} f(\underline{D} \mid \mathbf{M}_j)P(\mathbf{M}_j)} \tag{3.70}$$

where $P(\mathbf{M}_i)$ denotes the prior probability for $\mathbf{M}_i$. If a (non-informative) uniform prior is adopted over $\mathbf{M}_1$, $\mathbf{M}_2$, …, $\mathbf{M}_p$, that is, $P(\mathbf{M}_1) = P(\mathbf{M}_2) = \ldots = P(\mathbf{M}_p) = 1/p$, the posterior probability for each model is simply

$$P(\mathbf{M}_i \mid \underline{D}) = \frac{f(\underline{D} \mid \mathbf{M}_i)}{\sum_{j=1}^{p} f(\underline{D} \mid \mathbf{M}_j)} \tag{3.71}$$

Although the model selection can be conducted for any Bayesian problem with competing models, we will only demonstrate the model selection for Bayesian problems with spatial variability in this section.

EXAMPLE 3.14: Bayesian model selection for different trends of spatial variability data

The same spatial data $\underline{D} = \{\xi(z_1), \xi(z_2), \ldots, \xi(z_m)\}$ in Example 3.13 is analyzed here, but now consider three different trend models:

$\mathbf{M}_1$ (constant trend): $t(z) = a$, where $a \sim \text{unif}[0, 500]$

$\mathbf{M}_2$ (linear trend): $t(z) = a + b \times z$, where $a \sim \text{unif}[0, 500]$, $b \sim \text{unif}[-20, 20]$

$\mathbf{M}_3$ (quadratic trend): $t(z) = a + b \times z + c \times z^2$, where $a \sim \text{unif}[0, 500]$, $b \sim \text{unif}[-20, 20]$, $c \sim \text{unif}[-3, 3]$

The models for the spatial variability part (namely, the WM auto-correlation model and the prior PDFs for σ, δ, and ν) remain the same as in Example 3.13. Note that $\mathbf{M}_1$ (constant trend) is wrong for the data $\underline{D}$ because the actual trend is linear. In principle, both $\mathbf{M}_2$ (linear trend) and $\mathbf{M}_3$ (quadratic trend) are correct (a linear trend is a special case of $\mathbf{M}_3$ by setting c = 0), but $\mathbf{M}_3$ has a redundant part $c \times z^2$.

Because the TMCMC algorithm can estimate the model evidence $f(\underline{D}|\mathbf{M})$ as a by-product, the TMCMC algorithm is executed three times by adopting $\mathbf{M}_1$, $\mathbf{M}_2$, and $\mathbf{M}_3$. The logarithms of the resulting model evidence and posterior probabilities (assuming uniform priors $P(\mathbf{M}_1) = P(\mathbf{M}_2) = P(\mathbf{M}_3) = 1/3$) are as follows:

$$\ln[f(\underline{D}|\mathbf{M}_1)] = -704.9, P(\mathbf{M}_1|\underline{D}) = 0$$
$$\ln[f(\underline{D}|\mathbf{M}_2)] = -695.2, P(\mathbf{M}_2|\underline{D}) = 0.91$$
$$\ln[f(\underline{D}|\mathbf{M}_3)] = -697.6, P(\mathbf{M}_3|\underline{D}) = 0.09$$

which suggests that $\mathbf{M}_2$ is the most probable model given the spatial data $\underline{D}$.

EXAMPLE 3.15: Bayesian model selection for different auto-correlation models of spatial variability data

The same spatial data $\underline{D} = \{\xi(z_1), \xi(z_2), \ldots, \xi(z_m)\}$ in Example 3.13 is analyzed here, but now consider three different auto-correlation models:

$\mathbf{M}_1$: $\rho(\Delta z)$ = single exponential = $\exp(-2|\Delta z|/\delta)$, where $\ln(\delta) \sim \text{unif}[\ln(0.01), \ln(10)]$

$\mathbf{M}_2$: $\rho(\Delta z)$ = squared exponential = $\exp(-\pi\Delta z^2/\delta^2)$, where $\ln(\delta) \sim \text{unif}[\ln(0.01), \ln(10)]$

$\mathbf{M}_3$: $\rho(\Delta z)$ = WM model, where $\ln(\delta) \sim \text{unif}[\ln(0.01), \ln(10)]$ and $\ln(\nu) \sim \text{unif}[\ln(0.1), \ln(3)]$

The models for the trend part (namely, the linear trend model and the prior PDFs for a and b) and σ ($\ln(\sigma) \sim \text{unif}[\ln(0.1), \ln(100)]$) remain the same as in Example 3.13. Note that $\mathbf{M}_2$ (squared exponential auto-correlation) is wrong for the data $\underline{D}$ because the squared exponential model produces very smooth $\underline{D}$ realizations, but the actual smoothness of $\underline{D}$ is low. In principle, both $\mathbf{M}_1$ (single exponential auto-correlation) and $\mathbf{M}_3$ (WM auto-correlation) are correct (the single exponential auto-correlation is a special case of WM by setting $\nu = 0.5$).

Same as Example 3.14, the TMCMC algorithm is adopted here, so the model evidence $f(\underline{D}|\mathbf{M})$ is estimated automatically. The logarithms of the resulting model evidence and posterior probabilities (assuming uniform priors $P(\mathbf{M}_1) = P(\mathbf{M}_2) = P(\mathbf{M}_3) = 1/3$) are

$$\ln[f(\underline{D}|\mathbf{M}_1)] = -697.3, P(\mathbf{M}_1|\underline{D}) = 0.11$$
$$\ln[f(\underline{D}|\mathbf{M}_2)] = -712.1, P(\mathbf{M}_2|\underline{D}) = 0$$
$$\ln[f(\underline{D}|\mathbf{M}_3)] = -695.2, P(\mathbf{M}_3|\underline{D}) = 0.89$$

which suggests that $\mathbf{M}_3$ is the most probable model given the spatial data $\underline{D}$.

Chapter 4

Geotechnical data and Bayesian modeling

In Chapter 3, synthetic data were analyzed to demonstrate the Bayesian parameter estimation and prediction. In the current chapter, real geotechnical data, including cross-correlated data and spatially correlated data, are analyzed by the Bayesian approach. It is shown that simple models are usually ineffective in capturing real data. In order to effectively capture real data, more sophisticated models are usually needed. For instance, a hierarchical Bayesian model (HBM) (Gelman and Hill 2006) is adopted to model the site-uniqueness behavior in cross-correlated data, and a Gaussian process regression (GPR) model (Rasmussen and Williams 2006) is adopted to model the complicated trend function in the spatially correlated data. The presentation of the Bayesian analysis starts from simple but ineffective models to sophisticated but effective models.

4.1 CROSS-CORRELATED DATA

In geotechnical design, design soil parameters, such as the undrained shear strength of a clay or the friction angle of a sand, typically require relatively sophisticated and expensive sampling and laboratory testing techniques. Test indices such as standard penetration test (SPT) blow count or cone penetration test (CPT) values can be adopted to correlate to design soil parameters. The correlation between various soil parameters is called the cross-correlation in this chapter. In the geotechnical literature, a cross-correlation equation is often called a transformation model (Phoon and Kulhawy 1999b). For instance, it is common to estimate the friction angle (ϕ') of sand based on its CPT cone tip resistance (q_c) through a transformation model derived from calibration data points obtained in the literature (e.g., Figure 4.1). Useful compilations of transformation models are available in the literature (e.g., Kulhawy and Mayne 1990; Mayne et al. 2001).

In this section, soil property data that contain two clay parameters, including the undrained shear strength (s_u) of a clay and its (corrected) CPT cone tip resistance (q_t), are demonstrated. Based on a dataset of (s_u, q_t), a bivariate Nataf model is first constructed, and the Bayesian parameter estimation and

DOI: 10.1201/9781003309765-4

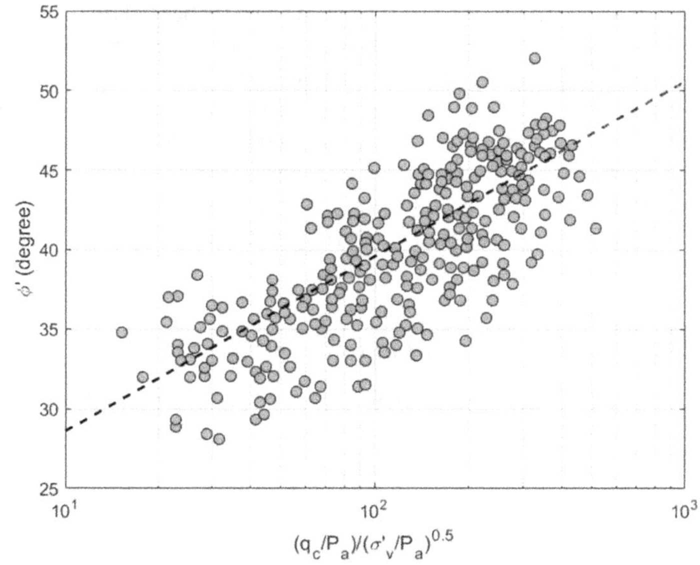

Figure 4.1 Transformation model between friction angle (ϕ′) of sand and cone tip resistance (q_c).

(Data source: Kulhawy and Mayne 1990).

prediction are conducted. The resulting model is a Bayesian transformation model that can predict the PDF of s_u of a clay given its q_t measurement. Three such Bayesian models will be presented, from simple to sophisticated models: (a) a generic model constructed based on a global (q_t, s_u) database; (b) a site-specific model constructed based on the (q_t, s_u) data at a Taipei (Taiwan) site; and (c) a quasi-site-specific model constructed based on both the global database and the Taipei-site data. Part (c) requires a hierarchical Bayesian model (HBM), which will be presented in detail in a later sub-section.

4.1.1 Generic model

Consider a clay database that contains 716 q_t vs. s_u records from 72 sites worldwide. This bivariate database is extracted from the CLAY/10/7490 database previously published by the author of this book (Ching and Phoon 2014). The data are normalized into $(q_t-\sigma_v)/\sigma'_v$ and s_u/σ'_v, where σ_v and σ'_v are the vertical total and effective stresses, respectively, for each record; the term $(q_t-\sigma_v)$ is the net cone tip resistance. The former normalized parameter $(q_t-\sigma_v)/\sigma'_v$ is further denoted by q_{t1} in this section. Empirically, it has been found that there is a positive correlation between $(q_t-\sigma_v)$ and s_u, so it is common to estimate the s_u of a clay based on its $(q_t-\sigma_v)$ information. In this section, we will construct a Bayesian transformation model between q_{t1} and s_u/σ'_v based on the database.

Figure 4.2a shows the q_{t1} vs. s_u/σ'_v data in the database. The histograms for $Y_1 = \ln(q_{t1})$ and $Y_2 = \ln(s_u/\sigma'_v)$ are shown in Figure 4.3. The logarithms are taken because q_{t1} and s_u/σ'_v are non-negative. We will construct a bivariate Nataf model (see Section 2.3.2) for the Y_1–Y_2 data. In other words, we first convert both (Y_1, Y_2) data to standard normal data, denoted by (X_1, X_2), as shown in Figure 4.2b. Then, we construct the bivariate normal PDF of (X_1, X_2) by Bayesian parameter estimation. Finally, the prediction of s_u for a future clay case based on its q_{t1} information will be achieved by Bayesian prediction. All procedures follow Chapter 3, but now real data are considered.

4.1.1.1 Conversion to standard normal variables

Due to its flexibility, the Johnson system of distributions (Johnson 1949) is adopted to model the marginal PDFs of (Y_1, Y_2). The Johnson system of distributions has been discussed in Section 2.2.4. Slifker and Shapiro (1980) proposed an elegant selection and parameter estimation approach for the Johnson system of distributions using percentiles:

1. Choose a number $z > 0$. We assume $z = 0.7$, as recommended in Slifker and Shapiro (1980).
2. Compute the sample fractiles of Y_i data corresponding to $-3z$, $-z$, z, $3z$ using the standard normal cumulative distribution function. For $z = 0.7$, the four sample fractiles are p_a-, p_b-, p_c-, and p_d-fractiles, where $p_a = \Phi(-2.1) = 0.018$, $p_b = \Phi(-0.7) = 0.242$, $p_c = \Phi(0.7) = 0.758$, and $p_d = \Phi(2.1) = 0.982$. Therefore, the four fractiles of Y_i are y_a = 0.018 sample fractile, y_b = 0.242 sample fractile, y_c = 0.758 sample fractile, and y_d = 0.982 sample fractile. In MATLAB command, y_i = fractile($\underline{Y}_i$,100*p_i), where the $\underline{Y}_i$ vector contains all Y_i data points.
3. Three parameters are computed from y_a, y_b, y_c, and y_d: $m = y_d - y_c$, $n = y_b - y_a$, $p = y_c - y_b$.
4. Finally, identify the Johnson member as SU if $mn/p^2 > 1$, SB if $mn/p^2 < 1$, and SL if $mn/p^2 = 1$.

Once the distribution type has been identified (SU, SB, or SL), the distribution parameters can be computed as follows:

For SU,

$$
\begin{aligned}
&\eta = 2z/\cosh^{-1}\left[0.5(m/p + n/p)\right] && \eta > 0 \\
&\gamma = \eta \times \sinh^{-1}\left\{(n/p - m/p)/\left[2(D-1)^{0.5}\right]\right\} && \\
&\lambda = 2p(D-1)^{0.5}/\left[(m/p + n/p - 2)(m/p + n/p + 2)^{0.5}\right] && \lambda > 0 \\
&\varepsilon = (y_b + y_c)/2 + p(n/p - m/p)/\left[2(m/p + n/p - 2)\right] &&
\end{aligned}
\tag{4.1}
$$

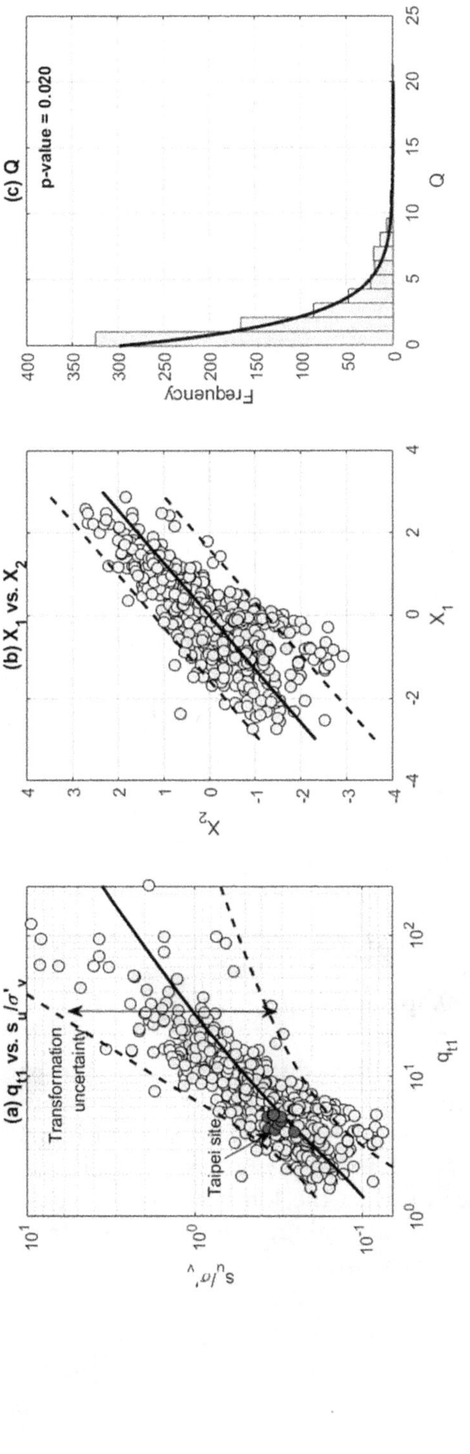

Figure 4.2 q_{t1} vs. s_u/σ'_v data in the soil database.

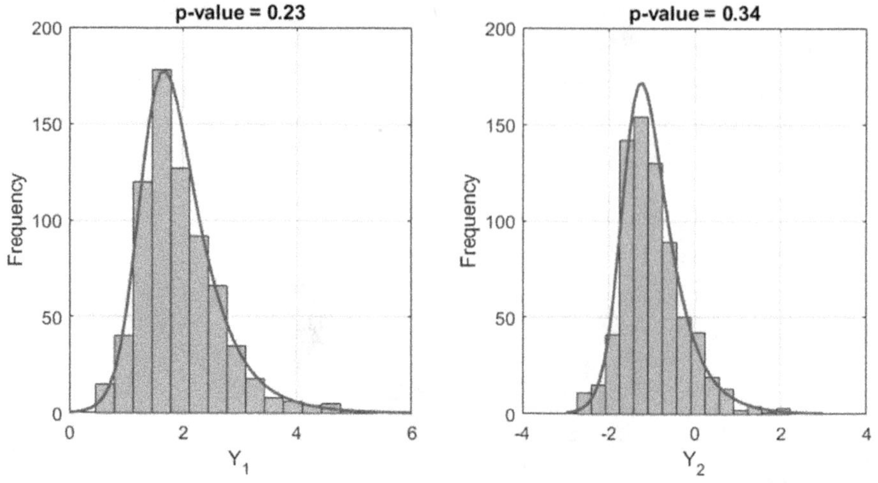

Figure 4.3 Histograms of Y_1 and Y_2.

For SB,

$$
\begin{aligned}
&\eta = z/\cosh^{-1}\left\{0.5\left[(1+p/m)(1+p/n)\right]^{0.5}\right\} && \eta > 0 \\
&\gamma = \eta \times \sinh^{-1}\left\{(p/n - p/m)\left[(1+p/m)(1+p/n) - 4\right]^{0.5} \Big/ \left[2\left(D^{-1}-1\right)\right]\right\} && \\
&\lambda = p\left\{\left[(1+p/m)(1+p/n) - 2\right]^2 - 4\right\}^{0.5} \Big/ \left(D^{-1}-1\right) && \lambda > 0 \\
&\varepsilon = \left(y_b + y_c\right)/2 - a_y/2 + p(p/n - p/m) \Big/ \left[2\left(D^{-1}-1\right)\right] && (4.2)
\end{aligned}
$$

For SL,

$$
\begin{aligned}
&\eta = 2z/\ln(m/p) \\
&\gamma = \eta \times \ln\left\{(m/p - 1) \Big/ \left[p(m/p)^{0.5}\right]\right\} \qquad (4.3) \\
&\varepsilon = \left(y_b + y_c\right)/2 - 0.5p(m/p + 1)/(m/p - 1)
\end{aligned}
$$

where $D = mn/p^2$. The resulting distribution types and parameters for (Y_1, Y_2) are shown in Table 4.1. The fitted marginal PDFs are shown in Figure 4.3. The p-values for the Kolmogorov–Smirnov test (Conover 1999) of the fitted PDFs are 0.23 and 0.34 for (Y_1, Y_2), respectively, suggesting that there is no strong evidence to reject the Johnson model. Then, the (Y_1, Y_2) data are converted to standard normal variables (X_1, X_2) data using the following transform for the SU distribution:

$$
X = \gamma + \eta \cdot \ln\left(\frac{Y-\varepsilon}{\lambda} + \sqrt{1 + \left(\frac{Y-\varepsilon}{\lambda}\right)^2}\right) \qquad (4.4)
$$

Table 4.1 Distribution type and distribution parameters for Y_1 and Y_2

Random variable	*Soil parameter*	*Distribution type*	*Distribution parameters*				*p-value*
			η	γ	λ	ε	
Y_1	$\ln(q_{t1})$	SU	1.793	−1.382	0.810	1.139	0.23
Y_2	$\ln(s_u/\sigma'_v)$	SU	1.721	−1.127	0.852	−1.680	0.34

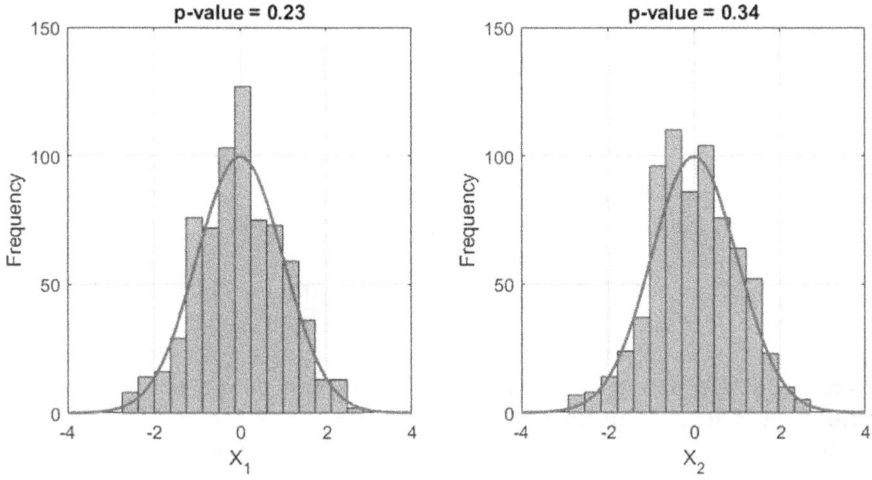

Figure 4.4 Histograms of X_1 and X_2.

Figure 4.4 shows the histograms of the converted (X_1, X_2) data, together with the standard normal PDF. Figure 4.2b shows the bivariate (X_1, X_2) data.

For the Nataf model, the (X_1, X_2) data are assumed to follow the bivariate normal PDF N(μ,C). Note that this bivariate normality is an "assumption" made by the Nataf model. It may or may not hold. To verify this assumption, one can further the (X_1, X_2) data into the following chi-squared Q data:

$$Q = \begin{bmatrix} X_1 - \mu_{est,1} \\ X_2 - \mu_{est,2} \end{bmatrix}^T C_{est}^{-1} \begin{bmatrix} X_1 - \mu_{est,1} \\ X_2 - \mu_{est,2} \end{bmatrix} \tag{4.5}$$

where $\mu_{est,i}$ is the sample mean of the X_i data, and C_{est} is the estimated covariance matrix of the (X_1, X_2) data. The histogram of the Q data is shown in Figure 4.2c. If the (X_1, X_2) data follow N(μ,C), the Q data follow the chi-squared PDF with a degree of freedom $\nu = 2$. The chi-squared PDF with $\nu = 2$ is also shown in Figure 4.2c for comparison. The Kolmogorov–Smirnov test for chi-squared PDF indicates that the p-value is 0.020, suggesting that there is sufficient evidence (with a significance level = 0.05) to reject the hypothesis that the Q data follow the chi-squared PDF, or equivalently to reject the bivariate normal PDF hypothesis for the (X_1, X_2) data. In principle, the Johnson–Nataf model is rejected. However, the following analysis still adopts the Johnson–Nataf model for the purpose of demonstration.

4.1.1.2 Bayesian parameter estimation

Given the independent and identically distributed (i.i.d.) generic data $\mathbf{X}$ = {$\underline{X}^{(1)}$, $\underline{X}^{(2)}$, ..., $\underline{X}^{(m)}$} (m = 716) in Figure 4.2b, the purpose is to estimate underlying $\underline{\mu}$ and $\mathbf{C}$. The assumed model is $\mathbf{M}$ = the bivariate normal PDF: $\underline{X} \sim N(\underline{\mu},\mathbf{C})$, where $\underline{\mu} = [\mu_1\ \mu_2]^T$ and $\mathbf{C} = [\sigma_1^2\ \rho_{12}\sigma_1\sigma_2; \rho_{12}\sigma_1\sigma_2\ \sigma_2^2]$ are both unknown. For brevity, the condition on $\mathbf{M}$ will be omitted in this chapter whenever the context is clear. We adopt the Gibbs sampler (GS) algorithm with T = 11,000 and T_b = 1000. Refer to Example 3.11 for the procedure of the GS algorithm for the bivariate normal PDF model. The following (roughly) non-informative conjugate priors are adopted: $\underline{\mu} \sim N(\underline{\mu}_0, \mathbf{C}_0)$ with $\underline{\mu}_0 = (0\ 0)^T$ and $\mathbf{C}_0 = 10{,}000^2 \times \mathbf{I}_{2\times2}$, and $\mathbf{C} \sim IW(\mathbf{\Sigma}_0, \nu_0)$ with $\mathbf{\Sigma}_0 = \mathbf{I}_{2\times2}$ and ν_0 = n +1 = 3. In the end, T-T_b = 10,000 posterior samples {($\underline{\mu}_t$,$\mathbf{C}_t$): t = T_b+1, ..., T} ~ f($\underline{\mu}$,$\mathbf{C}$|$\mathbf{X}$) are drawn by the GS algorithm. For the GS algorithm, samples at nearby time steps are correlated. To reduce sample correlation and also to reduce subsequent computation, only one sample is collected for every Δt time steps, that is, samples are collected at time steps T_b+1, T_b+Δt+1, ..., T. There are in total N = (T-T_b)/Δt samples, and let us denote the resulting samples by {($\underline{\mu}_k$,$\mathbf{C}_k$): k = 1, ..., N}. Figure 4.5 demonstrates the posterior samples {($\underline{\mu}_k$,$\mathbf{C}_k$): k = 1, ..., N} in terms of (μ_1, σ_1, μ_2, σ_2, ρ_{12}). It is clear that ($\underline{\mu}$,$\mathbf{C}$) is quite certain (95% confidence region is small). This is because there are many data (m = 716) such that the statistical uncertainty for ($\underline{\mu}$,$\mathbf{C}$) is small. Note that the resulting ($\underline{\mu}$,$\mathbf{C}$) is for the "generic model" because the 716 cases are generic data from different sites worldwide.

4.1.1.3 Bayesian prediction for a new case

Now consider a new clay case. Suppose the q_{t1} and s_u/σ'_v values of this clay also follow the same generic model, that is, $\underline{X}' = [X'_1\ X'_2]^T \sim N(\underline{\mu},\mathbf{C})$. Suppose that the q_{t1} value of the clay is known (i.e., $X'_1 = x'_1$ is known), and it is desirable to predict its s_u/σ'_v value (i.e., X'_2 is unknown). The purpose of the Bayesian prediction can be achieved if we can sample $X'_2 \sim f(x'_2|x'_1,\mathbf{X})$.

First, note that if $[X'_1\ X'_2]^T \sim N(\underline{\mu},\mathbf{C})$ and if $\underline{\mu} = [\mu_1\ \mu_2]^T$ and $\mathbf{C} = [\sigma_1^2\ \rho_{12}\sigma_1\sigma_2; \rho_{12}\sigma_1\sigma_2\ \sigma_2^2]$ are both known, the conditional PDF $f(x'_2|x'_1,\underline{\mu},\mathbf{C})$ is still normal with the following mean and variance:

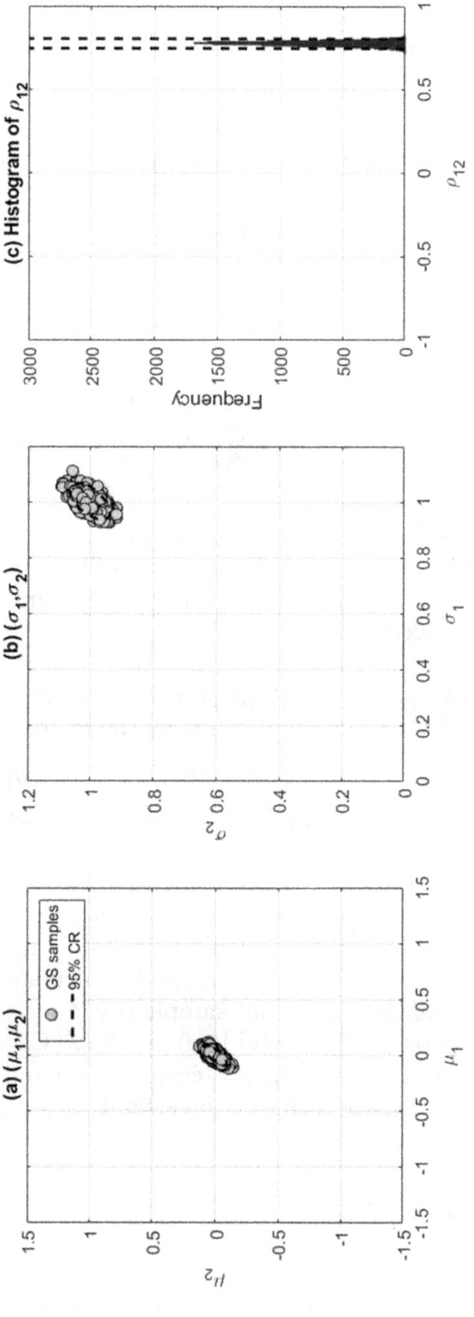

Figure 4.5 Posterior samples of ($\underline{\mu}$,**C**) (generic model).

$$
\begin{aligned}
E(X'_2 \mid X'_1, \underline{\mu}, C) &= E(X'_2) + COV(X'_2, X'_1)Var(X'_1)^{-1}\left[X'_1 - E(X'_1)\right] \\
&= \mu_2 + (\sigma_2\delta_{12} / \sigma_1) \cdot (X'_1 - \mu_1) \\
Var(X'_2 \mid X'_1, \underline{\mu}, C) &= Var(X'_2) - COV(X'_2, X'_1)Var(X'_1)^{-1} COV(X'_1, X'_2) \\
&= \sigma_2^2 - \sigma_2^2\rho_{12}^2 / \sigma_1^2
\end{aligned} \tag{4.6}
$$

According to the Total Probability Theorem and the Law of Large Numbers:

$$
\begin{aligned}
f(x'_2 \mid x'_1, \mathbf{X}) &= \int f(x'_2 \mid x'_1, \underline{\mu}, C, \mathbf{X}) f(\underline{\mu}, C \mid \mathbf{X}) d\underline{\mu} \cdot dC \\
&= \int f(x'_2 \mid x'_1, \underline{\mu}, C) f(\underline{\mu}, C \mid \mathbf{X}) d\underline{\mu} \cdot dC \\
&\approx \frac{1}{N}\sum_{k=1}^{N} f(x'_2 \mid x'_1, \underline{\mu}_k, C_k) \\
&= \frac{1}{N}\sum_{k=1}^{N} N\left(\mu_{2,k} + (\sigma_{2,k}\rho_{12,k}/\sigma_{1,k}) \cdot (x'_1 - \mu_{1,k}), \sigma_{2,k}^2 - \sigma_{2,k}^2\rho_{12,k}^2/\sigma_{1,k}^2\right)
\end{aligned} \tag{4.7}
$$

where $(\underline{\mu}_k, C_k)$ is the kth GS sample, and $(\mu_{1,k}, \sigma_{1,k}, \mu_{2,k}, \sigma_{2,k}, \rho_{12,k})$ can be extracted from $(\underline{\mu}_k, C_k)$. According to Equation (4.7), given the CPT value of the new clay case (i.e., $X'_1 = x'_1$), $X'_2 \sim f(x'_2|x'_1,\mathbf{X})$ can be then sampled using the following procedure:

1. Randomly draw $(\underline{\mu}, C)$ from $\{(\underline{\mu}_k, C_k): k = 1, ..., N\}$ and denote this sample by $(\hat{\underline{\mu}}, \hat{C})$ and their components by $(\hat{\mu}_1, \hat{\sigma}_1, \hat{\mu}_2, \hat{\sigma}_2, \hat{\rho}_{12})$.
2. Draw a sample $X'_2 \sim N\left(\hat{\mu}_2 + (\hat{\sigma}_2\hat{\rho}_{12} / \hat{\sigma}_1) \cdot (x'_1 - \hat{\mu}_1), \hat{\sigma}_2^2 - \hat{\sigma}_2^2\hat{\rho}_{12}^2 / \hat{\sigma}_1^2\right)$.
3. Cycle Steps 1 and 2 to obtain the desirable number of X'_2 samples.

A set of x'_1 values from −3 to 3 are investigated. For each x'_1 value, X'_2 samples $\sim f(x'_2|x'_1,\mathbf{X})$ are obtained using the above procedure. The posterior median estimate for X'_2 can be calculated as the sample 0.5-fractile of the X'_2 samples, and the posterior 95% CI can be calculated as the interval between the sample 0.025-fractile and sample 0.975-fractile. Figure 4.2b shows how the posterior median (solid line) and 95% CI (dashed lines) of X'_2 vary with the input x'_1 value. The posterior median and 95% CI can be mapped back to the Y'_2 space using the inverse-SU transform (the inverse equation of Equation 4.4):

$$
Y'_2 = \varepsilon_2 + \lambda_2 \cdot \sinh\left(\frac{X'_2 - \gamma_2}{\eta_2}\right) \tag{4.8}
$$

where $(\eta_2, \gamma_2, \lambda_2, \varepsilon_2)$ are the Johnson SU parameters for Y_2 (see Table 4.1). Figure 4.2a shows how the posterior median (solid line) and 95% CI (dashed lines) of Y'_2 varies with the input y'_1 value. The vertical distance between the dashed lines quantifies the magnitude of "transformation uncertainty". It is clear that the generic model has significant transformation uncertainty.

4.1.1.4 Bayesian prediction for a Taipei site

Now consider a clay layer at a Taipei site (Ou and Liao 1987). The CPT cone tip resistance (q_t) value versus depth for this clay layer is known and shown in Figure 4.6a. Based on the profiles of q_t, σ_v, and σ'_v, the $y'_1 = \ln(q_{t1})$ profile can be obtained and converted to x'_1 profile through the Johnson SU transform (e.g., Equation 4.4). Then, given the x'_1 value at each depth, the posterior median and 95% CI of X'_2 are obtained by the procedure above, and they are further converted back to the posterior median and 95% CI of $Y'_2 = \ln(s_u/\sigma'_v)$ through the inverse-SU transform (Equation 4.8). The posterior median and 95% CI profiles of s_u (shown in Figure 4.6b) can then be obtained by multiplying the profile of σ'_v with the posterior median and 95% CI profiles of s_u/σ'_v. Figure 4.6b shows the resulting posterior median

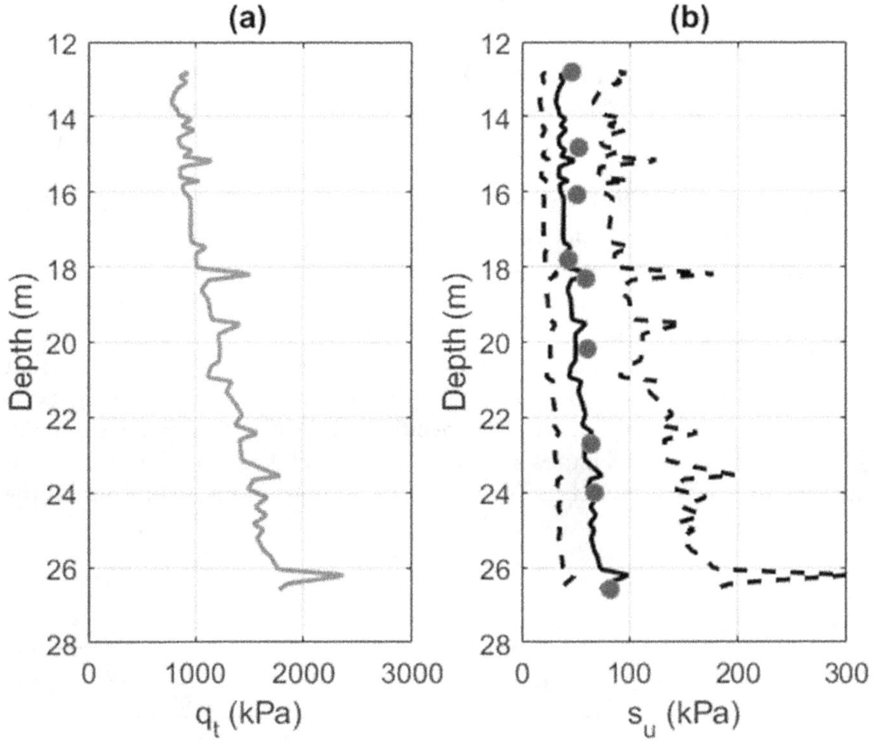

Figure 4.6 Results for the clay layer at a Taipei site (generic model). (a) q_t = data and (b) Prediction for s_u.

(solid line) and 95% CI (dashed lines) profiles of s_u. It is clear that the 95% CI is quite wide due to the significant transformation uncertainty of the generic model. Figure 4.6b also shows nine actual s_u data at this Taipei site for validation (the dots): all the actual s_u data fall within the 95% CI.

4.1.2 Site-specific model

The main reason that the generic model in Section 4.1.1 has a significant transformation uncertainty is that it is constructed based on generic data coming from various sites worldwide. For a particular site, there is usually a site-specific trend, and the site-specific transformation uncertainty is usually smaller than the generic transformation uncertainty. As a result, it is desirable to construct a site-specific transformation model. For the Taipei site, there are s_u data at nine depths (Figure 4.6b). The q_{t1} values for these nine depths are also known. Table 4.2 shows the q_{t1} vs. s_u/σ'_v data for these nine depths, and they are also shown as the dark dots in Figure 4.2a. These nine data are converted to $\underline{X}_s = (X_{s1}, X_{s2})$ data using the same Johnson parameters in Table 4.1 (the subscript "s" denotes "site-specific data").

4.1.2.1 Bayesian parameter estimation

Given the i.i.d. sites-specific data $\mathbf{X}_s = \{\underline{X}_s^{(1)}, \underline{X}_s^{(2)}, \ldots, \underline{X}_s^{(ms)}\}$ ($m_s = 9$), the purpose is to estimate the underlying site-specific mean vector ($\underline{\mu}_s$) and site-specific covariance matrix ($\mathbf{C}_s$). The assumed model is $\mathbf{M}$ = the bivariate normal PDF $N(\underline{\mu}_s, \mathbf{C}_s)$. We adopt the GS algorithm with the same parameters ($T = 11{,}000$, $T_b = 1000$, $\Delta t = 10$, and the non-informative priors). Figure 4.7 demonstrates the posterior samples $\{(\underline{\mu}_{s,k}, \mathbf{C}_{s,k})\text{: } k = 1, \ldots, N\}$ in terms of $(\mu_1, \sigma_1, \mu_2, \sigma_2, \rho_{12})$. The resulting $(\underline{\mu}_s, \mathbf{C}_s)$ samples are distributed as $f(\underline{\mu}_s, \mathbf{C}_s|\mathbf{X}_s)$: they represent the "site-specific model" because the nine cases are from the Taipei site. It is remarkable that the Taipei site has a site-specific mean vector different from the generic one (Figures 4.7a vs. 4.5a), which suggests that the Taipei site has a site-specific q_{t1} vs. s_u/σ'_v trend different from the generic trend. It is also clear that the site-specific standard deviations are smaller than the generic ones (Figures 4.7b vs. 4.5b), which suggests that the Taipei-site transformation model has smaller transformation uncertainty than the

Table 4.2 q_{t1} vs. s_u/σ'_v data for the Taipei site

Depth (m)	*12.8*	*14.8*	*16.1*	*17.8*	*18.3*	*20.2*	*22.7*	*24.0*	*26.6*
q_{t1}	5.17	4.22	4.12	4.03	5.27	4.53	4.76	5.12	5.32
s_u/σ'_v	0.367	0.365	0.332	0.252	0.340	0.319	0.305	0.304	0.337

Source: Ou and Liao 1987.

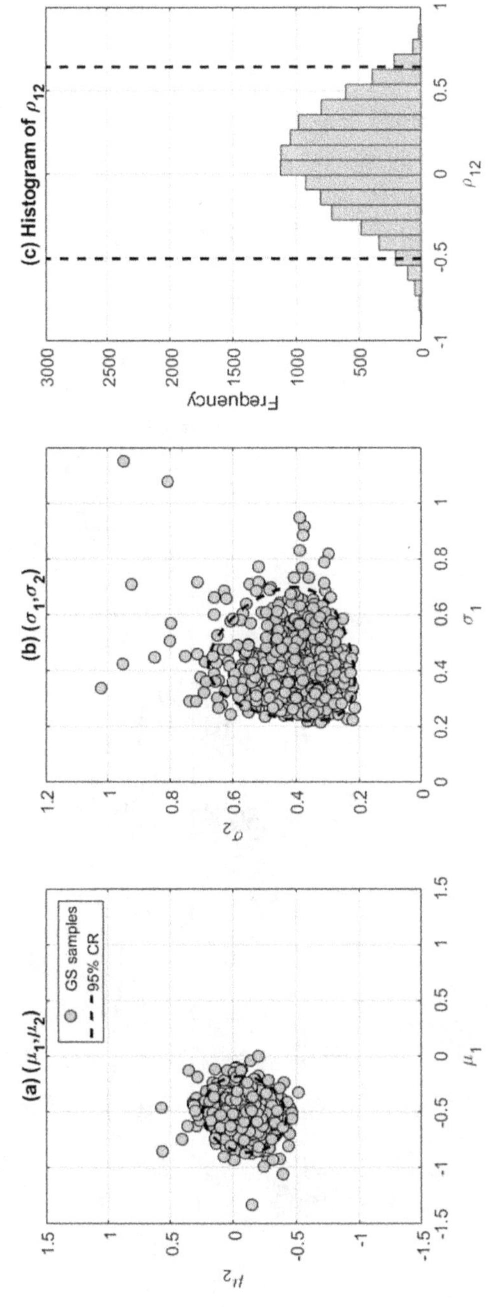

Figure 4.7 Posterior samples of ($\underline{\mathbf{\mu}}_s$,$\mathbf{C}_s$) (site-specific model).

generic model. However, the scatter of the (μ_s,C_s) samples is larger (95% confidence region is larger) than the scatter of the generic (μ,C) samples (Figures 4.7 vs. 4.5). This is because the site-specific data are sparse (m_s = 9) such that the statistical uncertainty for (μ_s,C_s) is significant.

4.1.2.2 Bayesian prediction

Now consider the depths at the Taipei site that are different from the nine depths. Recall that the q_t values at these depths are known (as shown in Figure 4.8a), and it is desirable to predict the s_u/σ'_v values at these depths. The same procedure for Bayesian prediction in the previous section can be adopted, but now the (μ_s,C_s) samples in Figure 4.7 for the site-specific model are adopted to obtain the posterior median and 95% CI profiles of s_u/σ'_v. The resulting posterior median and 95% CI profiles of s_u/σ'_v are shown in Figure 4.8b. The width of the 95% CI is still large due to the significant statistical uncertainty of the site-specific model.

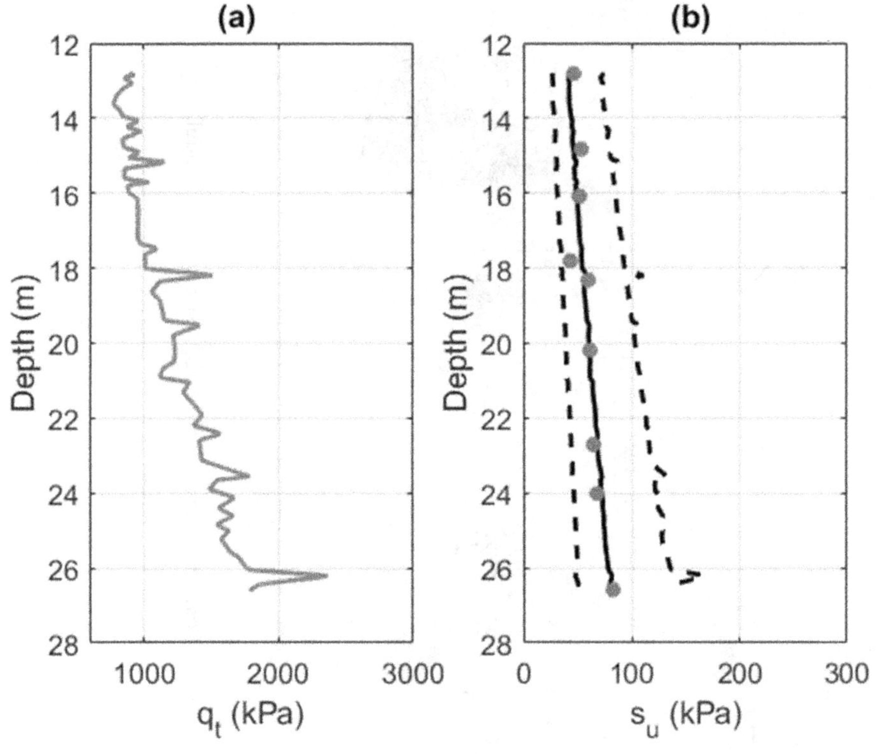

Figure 4.8 Results for the clay layer at a Taipei site (site-specific model). (a) q_t = data and (b) Prediction for s_u.

4.1.3 Quasi-site-specific model – hierarchical Bayesian modeling

Now we have seen a dilemma: the generic model has significant transformation uncertainty (due to pooling data from different sites together), but the site-specific model has significant statistical uncertainty (due to sparse site-specific data). The generic model ignores the "site uniqueness" by pooling all data from different sites together, whereas the site-specific model ignores the past experiences in the soil database by adopting the non-informative priors.

In this section, the HBM proposed by Ching et al. (2021a) is proposed to model the site uniqueness explicitly. This HBM can be trained by the soil database to learn the site-uniqueness characteristics in the database. The trained HBM can construct an "informative prior model" for the Taipei site. Then, this informative prior model is further updated by the Taipei-site data to obtain the posterior model. The resulting posterior model is "quasi-site-specific" because it is based on the prior model learned from the database and then updated by the Taipei-site data. It fuses two different sources of information (prior information from the database and the site-specific information from the Taipei site).

To illustrate the site-uniqueness characteristics in the soil database, Figure 4.9a shows the q_{t1} vs. s_u/σ'_v data in the database but now data from a particular site are shown as a particular color. For instance, the filled data points within the dashed circle in Figure 4.9a are from the same site. It is clear that these data follow a site-specific q_{t1} vs. s_u/σ'_v trend (site-specific transformation model) distinct from the generic trend (generic transformation model). Moreover, the scatter of the site-specific data around the site-specific trend (site-specific transformation uncertainty) is less than the scatter of the generic data around the generic trend (generic transformation uncertainty). This site uniqueness suggests that a site-specific transformation model should be more effective than the generic one. The purpose of the HBM is to learn the site-uniqueness characteristics in Figure 4.9a.

4.1.3.1 Hierarchical Bayesian modeling

The HBM proposed by Ching et al. (2021a) has the model structure in Figure 4.10. Recall that there are 72 sites in the database (the number of sites is denoted by $n_s = 72$). Suppose that for the ith site, there are j = 1, 2, ... m_i data points of (X_1, X_2). Let us denote the jth data at the ith site by $\underline{X}_i^{(j)} \in \mathbf{R}^{n\times 1}$ (n = 2 for the current example). For the ith site, $\underline{X}_i^{(j)}$ is assumed to follow the multivariate normal PDF $N(\underline{\mu}_i, C_i)$, where $\underline{\mu}_i \in \mathbf{R}^{n\times 1}$ and $C_i \in \mathbf{R}^{n\times n}$ are the site-specific mean vector and site-specific covariance matrix, respectively. The mean vectors of different sites in the database, that is, $\{\underline{\mu}_1, \underline{\mu}_2, \ldots, \underline{\mu}_{ns}\}$, are distinct but assumed to follow a common distribution governed by a (hyper) multivariate normal PDF $N(\underline{\mu}_0, C_0)$, where $\underline{\mu}_0 \in \mathbf{R}^{n\times 1}$

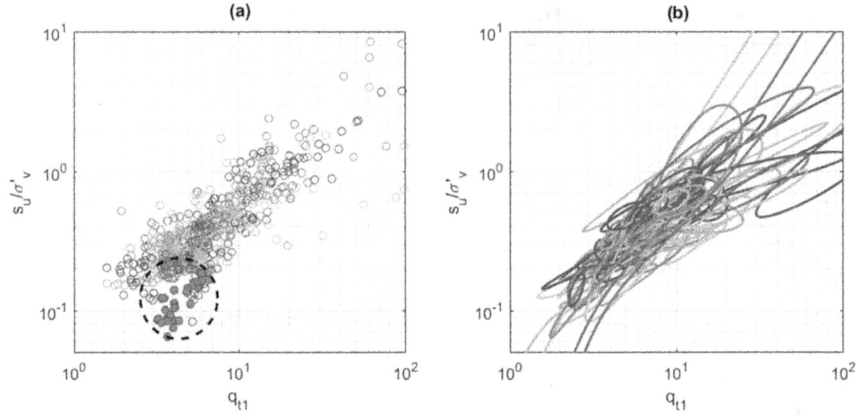

Figure 4.9 Site-uniqueness characteristics in the (a) q_{t1} vs. s_u/σ'_v data with site labels and (b) hypothetical sites simulated by trained HBM.

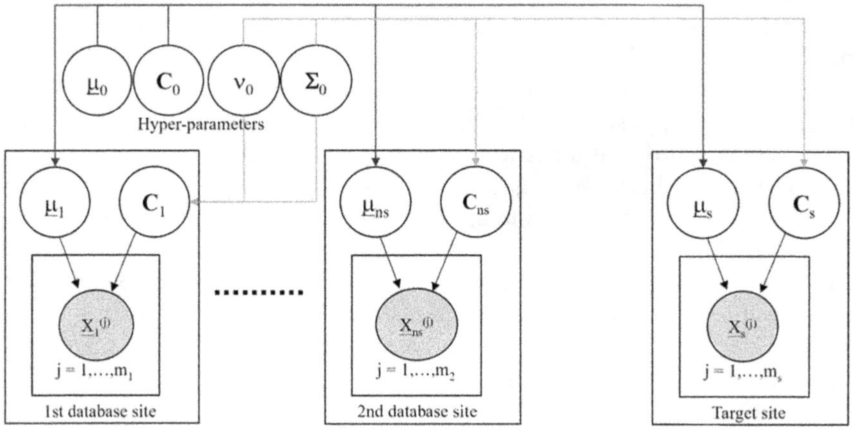

Figure 4.10 Model structure of HBM (modified with permission from American Society of Civil Engineers (Ching et al. 2021a)).

and $C_0 \in R^{n\times n}$ are the hyper-mean vector and hyper-covariance matrix for $\{\mu_1, \mu_2, \ldots, \mu_{ns}\}$, respectively. Similarly, the covariance matrices of different sites, that is, $\{C_1, C_2, \ldots, C_{ns}\}$, are distinct but assumed to follow a common distribution governed by a (hyper) inverse-Wishart PDF $IW(\boldsymbol{\Sigma}_0, \nu_0)$, where $\boldsymbol{\Sigma}_0 \in R^{n\times n}$ and $\nu_0 \in$ {the set of all integers} are the hyper-scale matrix and hyper-degree of freedom for $(C_1, C_2, \ldots, C_{ns})$, respectively. The parameters $\boldsymbol{\Theta} = \{\mu_0, C_0, \boldsymbol{\Sigma}_0, \nu_0\}$ are called the hyper-parameters. They govern the site-uniqueness characteristics, that is, the site-specific means $\{\mu_1, \mu_2, \ldots, \mu_{ns}\}$ and site-specific covariance matrices $\{C_1, C_2, \ldots, C_{ns}\}$. Let us denote $\{\mu_1, \mu_2, \ldots, \mu_{ns}\}$ by $\boldsymbol{\mu}$ and denote $\{C_1, C_2, \ldots, C_{ns}\}$ by $\mathbf{C}$. Let us further denote the database data by $\mathbf{X} = \{\underline{X}_i^{(j)}: i = 1, \ldots, n_s; j = 1, \ldots, m_j\}$. Note that in this section, the assumed model is $\mathbf{M}$ = the HBM. Again, the condition on $\mathbf{M}$ will be omitted whenever the context is clear.

The following PDFs are required to specify the joint distribution of all variables $(\mathbf{X},\boldsymbol{\mu},\mathbf{C},\boldsymbol{\Theta})$ in the HBM, that is, $f(\mathbf{X},\boldsymbol{\mu},\mathbf{C},\boldsymbol{\Theta})$. We describe the required PDFs from the top level (hyper-parameters) in Figure 4.10 to the bottom level ($\underline{X}_i^{(j)}$ data).

1. (Prior PDFs of $\underline{\mu}_0$, $\mathbf{C}_0$, $\boldsymbol{\Sigma}_0$, and ν_0): The top-level variables in Figure 4.10 are the hyper-parameters $\boldsymbol{\Theta} = \{\underline{\mu}_0,\mathbf{C}_0,\boldsymbol{\Sigma}_0,\nu_0\}$. In the Bayesian analysis, it is necessary to specify the prior PDFs for these top-level variables. It is desirable to adopt conjugate priors such that their posterior PDFs have analytical expressions.
   a. (Prior PDF of $\underline{\mu}_0$): The conjugate prior for $\underline{\mu}_0$ is a multivariate normal PDF, $N(\underline{\mu}_{\mu 0},\mathbf{C}_{\mu 0})$:

$$N\left(\underline{\mu}_{\mu 0},\mathbf{C}_{\mu 0}\right) = \left|\mathbf{C}_{\mu 0}\right|^{-\frac{1}{2}} \cdot (2\pi)^{-\frac{n}{2}} \cdot \exp\left[-\frac{1}{2}\left(\underline{\mu}_0 - \underline{\mu}_{\mu 0}\right)^T \mathbf{C}_{\mu 0}^{-1}\left(\underline{\mu}_0 - \underline{\mu}_{\mu 0}\right)\right] \quad (4.9)$$

where $\underline{\mu}_{\mu 0} \in \mathbf{R}^{n\times 1}$ is the mean vector and $\mathbf{C}_{\mu 0} \in \mathbf{R}^{n\times n}$ is the covariance matrix. To make this prior non-informative, we adopt $\underline{\mu}_{\mu 0}$ to be a zero vector and $\mathbf{C}_{\mu 0}$ to be a diagonal matrix with larger diagonals (e.g., 10^4).
   b. (Prior PDF of $\mathbf{C}_0$): The conjugate prior for $\mathbf{C}_0$ is an inverse-Wishart PDF, $IW(\boldsymbol{\Sigma}_{C0},\nu_{C0})$:

$$IW\left(\boldsymbol{\Sigma}_{C_0},\nu_{C_0}\right) = \frac{\left|\boldsymbol{\Sigma}_{C_0}\right|^{\nu_{C0}/2}}{2^{n\times\nu_{C0}/2}\cdot\Gamma_n\left(\nu_{C_0}/2\right)} \cdot \left|\mathbf{C}_0\right|^{-\frac{\nu_{C0}+n+1}{2}} \cdot \exp\left[-\frac{1}{2}\mathrm{tr}\left(\boldsymbol{\Sigma}_{C_0}\times\mathbf{C}_0^{-1}\right)\right] \quad (4.10)$$

where $\Gamma_n(.)$ is the multivariate Gamma function with dimension = d; $\boldsymbol{\Sigma}_{C0}$ is the scale matrix; ν_{C0} is the degree of freedom. It is not straightforward to make an inverse-Wishart PDF non-informative. The setting $\boldsymbol{\Sigma}_{C0} = \mathbf{I}_{n\times n}$ and $\nu_{C0} = n + 1$ is only weakly informative. To achieve a more non-informative IW PDF, Huang and Wand (2013) proposed that ν_{C0} = n+1 and $\boldsymbol{\Sigma}_{C0} = 4\times\mathrm{diag}(1/a_1, \ldots, 1/a_n)$ (diag(.) means a diagonal matrix), and a_k is random and distributed as an inverse-Gamma PDF, $IG(\alpha,\beta)$:

$$IG(\alpha,\beta) = \frac{\beta^\alpha}{\Gamma(\alpha)} \cdot a_k^{-\alpha-1} \cdot \exp\left(-\frac{\beta}{a_k}\right) \quad (4.11)$$

where α is the shape parameter and β is the scale parameter. Huang and Wand (2013) showed that α = 0.5 and β = a small number (e.g., 10^{-4}) make $IW(\mathbf{\Sigma}_{C0},\nu_{C0})$ more non-informative than simply adopting $\mathbf{\Sigma}_{C0} = \mathbf{I}_{n\times n}$ and $\nu_{C0} = n + 1$.

c. (Prior PDF of $\mathbf{\Sigma}_0$): The conjugate prior for $\mathbf{\Sigma}_0$ is a Wishart PDF:

$$W\left(\Psi_{\Sigma_0},\lambda_{\Sigma_0}\right)=\frac{\left|\Psi_{\Sigma_0}\right|^{-\lambda_{\Sigma 0}/2}}{2^{n\times\lambda_{\Sigma 0}/2}\cdot\Gamma_n\left(\lambda_{\Sigma_0}/2\right)}\cdot\left|\Sigma_0\right|^{\frac{\lambda_{\Sigma 0}-n-1}{2}}\cdot\exp\left[-\frac{1}{2}\mathrm{tr}\left(\Sigma_0\times\Psi_{\Sigma_0}^{-1}\right)\right] \tag{4.12}$$

where $\mathbf{\Psi}_{\Sigma 0}$ is the scale matrix; $\lambda_{\Sigma 0}$ is the degree of freedom. Chung et al. (2015) showed that if $\lambda_{\Sigma 0} = n+2$ and $\mathbf{\Psi}_{\Sigma 0}$ is a diagonal matrix with large diagonals (e.g., 10^4), $W(\mathbf{\Psi}_{\Sigma 0},\lambda_{\Sigma 0})$ is roughly non-informative.

d. (Prior distribution ν_0): ν_0 is an integer ranging from d to infinity. To make the prior distribution of ν_0 roughly non-informative, a uniform distribution between n and a large upper bound (e.g., 1000) is adopted.

2. (Conditional PDF of $\underline{\mu}_i$ and C_i): The middle-level variables in Figure 4.10 are $\boldsymbol{\mu} = \{\underline{\mu}_1, \underline{\mu}_2, \ldots, \underline{\mu}_{ns}\}$ and $\mathbf{C} = \{C_1, C_2, \ldots, C_{ns}\}$. Their "parents" are the hyper-parameters $\mathbf{\Theta}$. In Bayesian analysis, it is necessary to specify the conditional PDFs of $\underline{\mu}_i$ and C_i conditioning on their parents $\mathbf{\Theta}$:

a. (Conditional PDF of $\underline{\mu}_i$): $\underline{\mu}_i$ is assumed to follow $N(\underline{\mu}_0,C_0)$:

$$N\left(\underline{\mu}_0,C_0\right)=\left|C_0\right|^{-\frac{1}{2}}\cdot(2\pi)^{-\frac{n}{2}}\cdot\exp\left[-\frac{1}{2}\left(\underline{\mu}_i-\underline{\mu}_0\right)^T C_0^{-1}\left(\underline{\mu}_i-\underline{\mu}_0\right)\right] \tag{4.13}$$

b. (Conditional PDF of C_i): C_i is assumed to follow $IW(\mathbf{\Sigma}_0,\nu_0)$:

$$IW\left(\Sigma_0,\nu_0\right)=\frac{\left|\Sigma_0\right|^{\nu_0/2}}{2^{n\times\nu_0/2}\cdot\Gamma_n\left(\nu_0/2\right)}\cdot\left|C_i\right|^{-\frac{\nu_0+n+1}{2}}\cdot\exp\left[-\frac{1}{2}\mathrm{tr}\left(\Sigma_0\times C_i^{-1}\right)\right] \tag{4.14}$$

3. (Conditional PDF of $\underline{X}_i^{(j)}$): The bottom-level variables in Figure 4.10 are $\mathbf{X} = \{\underline{X}_i^{(j)}: i = 1, \ldots, n_s; j = 1, \ldots, m_i\}$. Their parents are $\boldsymbol{\mu} = \{\underline{\mu}_1, \underline{\mu}_2, \ldots, \underline{\mu}_{ns}\}$ and $\mathbf{C} = \{C_1, C_2, \ldots, C_{ns}\}$. In Bayesian analysis, it is necessary to specify the conditional PDF $\underline{X}_i^{(j)}$ conditioning on their parents $\underline{\mu}_i$ and C_i: $\underline{X}_i^{(j)}$ is assumed to follow $N(\underline{\mu}_i,C_i)$:

$$N\left(\underline{\mu}_i,C_i\right)=\left|C_i\right|^{-\frac{1}{2}}\cdot(2\pi)^{-\frac{n}{2}}\cdot\exp\left[-\frac{1}{2}\left(\underline{x}_i^{(j)}-\underline{\mu}_i\right)^T C_i^{-1}\left(\underline{x}_i^{(j)}-\underline{\mu}_i\right)\right] \tag{4.15}$$

Given $\underline{\mu}_i$ and C_i, we further assume $\underline{X}_i^{(j)}$ and $\underline{X}_i^{(k)}$ to be independent.

The variables in the proposed HBM include $\mathbf{X}$, $\boldsymbol{\mu}$, C, $\mathbf{\Theta}$, and $\underline{a} = \{a_1, a_2, \ldots, a_n\}$.

4.1.3.2 *Estimation of the hyper-parameters – construction of an informative prior model*

Recall that the hyper-parameters **Θ** govern the site-uniqueness characteristics in the soil database. Therefore, the purpose here is to estimate the hyper-parameters **Θ** based on the database **X** = $\{\underline{X}_i^{(j)}: i = 1, \ldots, n_s; j = 1, \ldots, m_i\}$. Once the hyper-parameters **Θ** are estimated, an informative prior model can then be constructed. The estimation of **Θ** can be achieved by drawing **Θ** samples ~ f(**Θ**|**X**). The assumed model is **M** = the HBM. To draw **Θ** samples ~ f(**Θ**|**X**), the Gibbs sampler (GS) algorithm is adopted to first draw ($\boldsymbol{\mu}$,C,**Θ**,$\underline{a}$) samples ~ f($\boldsymbol{\mu}$,C,**Θ**,$\underline{a}$|**X**). If we only keep the samples of the hyper-parameters **Θ**, these samples are distributed as f(**Θ**|**X**). The ($\boldsymbol{\mu}$,C,**Θ**,$\underline{a}$) samples ~ f($\boldsymbol{\mu}$,C,**Θ**,$\underline{a}$|**X**) can be drawn by the following GS algorithm:

1. Initialize ($\boldsymbol{\mu}$,C,**Θ**,$\underline{a}$) samples at arbitrary values.
2. (For i = 1, …, n_s): Draw $\underline{\mu}_i$ sample ~ $f(\underline{\mu}_i|\boldsymbol{\mu}_{\backslash i},C,\boldsymbol{\Theta},\underline{a})$, where $\boldsymbol{\mu}_{\backslash i}$ denotes $\{\underline{\mu}_1, \ldots, \underline{\mu}_{i-1}, \underline{\mu}_{i+1}, \ldots, \underline{\mu}_{ns}\}$.
3. (For i = 1, …, n_s): Draw C_i sample ~ $f(C_i|\boldsymbol{\mu},C_{\backslash i},\boldsymbol{\Theta},\underline{a})$, where $C_{\backslash i}$ denotes $\{C_1, \ldots, C_{i-1}, C_{i+1}, \ldots, C_{ns}\}$.
4. Draw $\underline{\mu}_0$ sample ~ $f(\underline{\mu}_0|\boldsymbol{\mu},C,C_0,\boldsymbol{\Sigma}_0,\nu_0,\underline{a})$.
5. Draw C_0 sample ~ $f(C_0|\boldsymbol{\mu},C,\underline{\mu}_0,\boldsymbol{\Sigma}_0,\nu_0,\underline{a})$.
6. Draw $\boldsymbol{\Sigma}_0$ sample ~ $f(\boldsymbol{\Sigma}_0|\boldsymbol{\mu},C,\underline{\mu}_0,C_0,\nu_0,\underline{a})$.
7. Draw ν_0 sample ~ $p(\nu_0|\boldsymbol{\mu},C,\underline{\mu}_0,C_0,\boldsymbol{\Sigma}_0,\underline{a})$, where p(.) denotes a discrete distribution.
8. (For k = 1, …, n): Draw a_k sample ~ $f(a_k|\boldsymbol{\mu},C,\boldsymbol{\Theta},\underline{a}_{\backslash k})$, where $\underline{a}_{\backslash k}$ denotes $(a_1, \ldots, a_{k-1}, a_{k+1}, \ldots, a_n)$.
9. Cycles 2–8 for T time steps to obtain T samples for ($\boldsymbol{\mu}$,C,**Θ**,$\underline{a}$).

The GS starts with the initial ($\boldsymbol{\mu}$,C,**Θ**,$\underline{a}$) and then sequentially draws samples from the fully conditional PDFs by conditioning on the latest parameter values (all parameters in the condition are fixed at their current sampled values). Due to the use of conjugate priors, it can be shown that many of the above fully conditional PDFs have the analytical expressions:

$$f(\underline{\mu}_i \mid \mathbf{X},\boldsymbol{\mu}_{\backslash i},C,\boldsymbol{\Theta},\underline{a}) = f(\underline{\mu}_i \mid \mathbf{X}_i,C_i)$$
$$= N\left\{\left(C_0^{-1}+m_iC_i^{-1}\right)^{-1}\left(C_0^{-1}\underline{\mu}_0 + C_i^{-1}\sum_{j=1}^{m_i}\underline{X}_i^{(j)}\right),\left(C_0^{-1}+m_iC_i^{-1}\right)^{-1}\right\} \tag{4.16}$$

$$f(C_i \mid \mathbf{X},\boldsymbol{\mu},C_{\backslash i},\boldsymbol{\Theta},\underline{a}) = f(C_i \mid \mathbf{X}_i,\underline{\mu}_i)$$
$$= IW\left\{\boldsymbol{\Sigma}_0 + \sum_{j=1}^{m_i}\left(\underline{X}_i^{(j)}-\underline{\mu}_i\right)\left(\underline{X}_i^{(j)}-\underline{\mu}_i\right)^T, m_i+\nu_0\right\} \tag{4.17}$$

$$f(\underline{\mu}_0 \mid \mathbf{X},\boldsymbol{\mu},\mathbf{C},\mathbf{C}_0,\boldsymbol{\Sigma}_0,\nu_0,\underline{a}) = f(\underline{\mu}_0 \mid \boldsymbol{\mu},\mathbf{C}_0)$$
$$= N\left\{\left(C_{\mu 0}^{-1} + n_s C_0^{-1}\right)^{-1}\left(C_{\mu 0}^{-1}\underline{\mu}_{\mu 0} + C_0^{-1}\sum_{i=1}^{n_s}\underline{\mu}_i\right), \left(C_{\mu 0}^{-1} + n_s C_0^{-1}\right)^{-1}\right\} \tag{4.18}$$

$$f(C_0 \mid \mathbf{X},\boldsymbol{\mu},\mathbf{C},\underline{\mu}_0,\boldsymbol{\Sigma}_0,\nu_0,\underline{a}) = f(C_0 \mid \boldsymbol{\mu},\underline{\mu}_0)$$
$$= IW\left\{\Sigma_{C_0} + \sum_{i=1}^{n_s}\left(\underline{\mu}_i - \underline{\mu}_0\right)\left(\underline{\mu}_i - \underline{\mu}_0\right)^T, n_s + \nu_{C0}\right\} \tag{4.19}$$

$$f(\boldsymbol{\Sigma}_0 \mid \mathbf{X},\boldsymbol{\mu},\mathbf{C},\underline{\mu}_0,C_0,\nu_0,\underline{a}) = f(\boldsymbol{\Sigma}_0 \mid \mathbf{C},\nu_0)$$
$$= W\left\{\left(\Psi_{\Sigma 0}^{-1} + \sum_{i=1}^{n_s} C_i^{-1}\right)^{-1}, n_s\nu_0 + \lambda_{\Sigma 0}\right\} \tag{4.20}$$

where $\mathbf{X}_i$ denotes $\{\underline{X}_i^{(j)}: j = 1, \ldots, m_i\}$. Note that in the above fully conditional PDFs, there are conditional independences, so some variables can be dropped out of the condition. Let us consider Equation (4.18) as an example: given $\{\boldsymbol{\mu},C_0\}$, $\underline{\mu}_0$ is independent of $\{\mathbf{X},\mathbf{C},\boldsymbol{\Sigma}_0,\nu_0,\underline{a}\}$, so the fully conditional PDF $f(\underline{\mu}_0|\mathbf{X},\boldsymbol{\mu},\mathbf{C},C_0,\boldsymbol{\Sigma}_0,\nu_0,\underline{a})$ is reduced to $f(\underline{\mu}_0|\boldsymbol{\mu},C_0)$. For this fully conditional PDF $f(\underline{\mu}_0|\boldsymbol{\mu},C_0)$, the unknown parameter is $\underline{\mu}_0$, the prior PDF is $f(\underline{\mu}_0|C_0) = f(\underline{\mu}_0) = N(\underline{\mu}_{\mu 0},C\underline{\mu}_0)$, and the "observations" are $\boldsymbol{\mu} = \{\underline{\mu}_1, \ldots, \underline{\mu}_{ns}\}$. Here, $\boldsymbol{\mu} = \{\underline{\mu}_1, \ldots, \underline{\mu}_{ns}\}$ are considered as "observations" because $\{\underline{\mu}_1, \ldots, \underline{\mu}_{ns}\}$ are fixed at their sampled values when the $\underline{\mu}_0$ sample ~ $f(\underline{\mu}_0|\boldsymbol{\mu},C_0)$ is drawn during the GS algorithm. This fully conditional PDF $f(\underline{\mu}_0|\boldsymbol{\mu},C_0)$ is the third scenario in Table 2.1 (multivariate normal with unknown mean vector but known covariance matrix). According to Table 2.1, this fully conditional (posterior) PDF $f(\underline{\mu}_0|\boldsymbol{\mu},C_0)$ is

$$f(\underline{\mu}_0 \mid \boldsymbol{\mu},C_0) = N\left\{\left(C_{\mu 0}^{-1} + n_s C_0^{-1}\right)^{-1}\left(C_0^{-1}\sum_{i=1}^{n_s}\underline{\mu}_i\right), \left(C_{\mu 0}^{-1} + n_s C_0^{-1}\right)^{-1}\right\} \tag{4.21}$$

Table 4.3 shows the list of the fully conditional PDFs of some parameters in the HBM.

The fully conditional PDF $f(a_k|\boldsymbol{\mu},\mathbf{C},\boldsymbol{\Theta},\underline{a}_{\backslash k})$ also has an analytical expression, but its derivation is more complicated. In general, all fully conditional PDFs can be derived by first writing down the complete multivariate PDF $f(\mathbf{X},\boldsymbol{\mu},\mathbf{C},\boldsymbol{\Theta},\underline{a})$ and then consider the parameter of interest as the only variable

Table 4.3 Fully conditional PDFs for some parameters in the HBM

Fully conditional PDF	*Unknown parameter*	*Observations*	*Likelihood*	*Conjugate prior PDF*	*Fully conditional PDF in conjugate form*
$f(\underline{\mu}_i \mid \mathbf{X}_i, \mathbf{C}_i)$	$\underline{\mu}_i$	$\mathbf{X}_i = \{\underline{X}_i^{(1)}, \ldots, \underline{X}_i^{(mi)}\}$, where $\underline{X}_i^{(j)} \sim N(\underline{\mu}_i, \mathbf{C}_i)$	Multivariate normal w/ known $\mathbf{C}_i$	$N(\underline{\mu}_0, \mathbf{C}_0)$	$N\left\{\left(\mathbf{C}_0^{-1} + m_i\mathbf{C}_i^{-1}\right)^{-1}\left(\mathbf{C}_0^{-1}\underline{\mu}_0 + \mathbf{C}_i^{-1}\sum_{j=1}^{m_i}\underline{X}_i^{(j)}\right), \left(\mathbf{C}_0^{-1} + m_i\mathbf{C}_i^{-1}\right)^{-1}\right\}$
$f(\mathbf{C}_i \mid \mathbf{X}_i, \underline{\mu}_i)$	$\mathbf{C}_i$	$\mathbf{X}_i = \{\underline{X}_i^{(1)}, \ldots, \underline{X}_i^{(mi)}\}$, where $\underline{X}_i^{(j)} \sim N(\underline{\mu}_i, \mathbf{C}_i)$	Multivariate normal w/ known $\underline{\mu}_i$	$IW(\mathbf{\Sigma}_0, \nu_0)$	$IW\left\{\Sigma_0 + \sum_{j=1}^{m_i}\left(\underline{X}_i^{(j)} - \underline{\mu}_i\right)\left(\underline{X}_i^{(j)} - \underline{\mu}_i\right)^T, m_i + \nu_0\right\}$
$f(\underline{\mu}_0 \mid \mathbf{\mu}, \mathbf{C}_0)$	$\underline{\mu}_0$	$\mathbf{\mu} = \{\underline{\mu}_1, \ldots, \underline{\mu}_{ns}\}$, where $\underline{\mu}_i \sim N(\underline{\mu}_0, \mathbf{C}_0)$	Multivariate normal w/ known $\mathbf{C}_0$	$N(\underline{\mu}_{\mu 0}, \mathbf{C}_{\mu 0})$	$N\left\{\left(\mathbf{C}_{\underline{\mu}0}^{-1} + n_s \cdot \mathbf{C}_0^{-1}\right)^{-1}\left(\mathbf{C}_0^{-1}\sum_{i=1}^{n_s}\underline{\mu}_i\right), \left(\mathbf{C}_{\underline{\mu}0}^{-1} + n_s \cdot \mathbf{C}_0^{-1}\right)^{-1}\right\}$
$f(\mathbf{C}_0 \mid \mathbf{\mu}, \underline{\mu}_0)$	$\mathbf{C}_0$	$\mathbf{\mu} = \{\underline{\mu}_1, \ldots, \underline{\mu}_{ns}\}$, where $\underline{\mu}_i \sim N(\underline{\mu}_0, \mathbf{C}_0)$	Multivariate normal w/ known $\underline{\mu}_0$	$IW(\mathbf{\Sigma}_{C0}, \nu_{C0})$	$IW\left\{\Sigma_{\mathbf{C}_0} + \sum_{i=1}^{n_s}\left(\underline{\mu}_i - \underline{\mu}_0\right)\left(\underline{\mu}_i - \underline{\mu}_0\right)^T, n_s + \nu_{C0}\right\}$
$f(\mathbf{\Sigma}_0 \mid \mathbf{C}, \nu_0)$	$\mathbf{\Sigma}_0$	$\mathbf{C} = \{\mathbf{C}_1, \ldots, \mathbf{C}_{ns}\}$, where $\mathbf{C}_i \sim IW(\mathbf{\Sigma}_0, \nu_0)$	Inverse-Wishart w/ known ν_0	$W(\mathbf{\Psi}_{\Sigma 0}, \lambda_{\Sigma 0})$	$W\left\{\left(\Psi_{\Sigma_0}^{-1} + \sum_{i=1}^{n_s}\mathbf{C}_i^{-1}\right)^{-1}, n_s\nu_0 + \lambda_{\Sigma_0}\right\}$

(other parameters are fixed constants). For the HBM, the complete multivariate PDF $f(\mathbf{X},\boldsymbol{\mu},\mathrm{C},\boldsymbol{\Theta},\underline{a})$ has the following expression:

$$
\begin{aligned}
& f(\mathbf{X},\boldsymbol{\mu},\mathrm{C},\boldsymbol{\Theta},\underline{a}) = f(\mathbf{X} \mid \boldsymbol{\mu},\mathrm{C}) \cdot f\left(\boldsymbol{\mu} \mid \underline{\mu}_0,\mathrm{C}_0\right) \cdot f(\mathrm{C} \mid \boldsymbol{\Sigma}_0,\nu_0) \cdot f\left(\underline{\mu}_0\right) \\
& \cdot f(\mathrm{C}_0) \cdot f(\boldsymbol{\Sigma}_0 | \underline{a}) \cdot p(\nu_0) \cdot f(\underline{a}) \\
& \propto \left(\prod_{i=1}^{n_s} \left|\mathrm{C}_i\right|^{-\frac{m_i+\nu_0+n+1}{2}}\right) \times \left|\mathrm{C}_0\right|^{-\frac{n_s+\nu_{C0}+n+1}{2}} \times \left|\boldsymbol{\Sigma}_0\right|^{\frac{n_s\nu_0+\lambda_{\Sigma 0}-n-1}{2}} \times \left|\begin{bmatrix} 4/a_1 & & \\ & \ddots & \\ & & 4/a_n \end{bmatrix}\right|^{\nu_{C0}/2} \\
& \times \frac{\prod_{k=1}^{n} a_k^{-\alpha-1}}{2^{n_s \times n \times \nu_0/2} \times \Gamma_n(\nu_0/2)^{n_s}} \\
& \times e^{-\frac{1}{2}\sum_{i=1}^{n_s}\sum_{j=1}^{m_i}\left(\underline{x}_i^{(j)}-\underline{\mu}_i\right)^T \mathrm{C}_i^{-1}\left(\underline{x}_i^{(j)}-\underline{\mu}_i\right)-\frac{1}{2}\sum_{i=1}^{n_s}\left(\underline{\mu}_i-\underline{\mu}_0\right)^T \mathrm{C}_0^{-1}\left(\underline{\mu}_i-\underline{\mu}_0\right)-\frac{1}{2}\left(\underline{\mu}_0-\underline{\mu}_{\mu 0}\right)^T \mathrm{C}_{\mu 0}^{-1}\left(\underline{\mu}_0-\underline{\mu}_{\mu 0}\right)-\frac{1}{2}\mathrm{tr}\left[\Sigma_0 \times \left(\Psi_{\Sigma 0}^{-1}+\sum_{i=1}^{n_s}\mathrm{C}_i^{-1}\right)\right] -\frac{1}{2}\mathrm{tr}\left(\begin{bmatrix} 4/a_1 & & \\ & \ddots & \\ & & 4/a_n \end{bmatrix} \times \mathrm{C}_0^{-1}\right)-\sum_{k=1}^{n}\frac{\beta}{a_k}}
\end{aligned}
\tag{4.22}
$$

For the fully conditional PDF $f(a_k|\boldsymbol{\mu},\mathrm{C},\boldsymbol{\Theta},\underline{a}_{\backslash k})$, the parameter of interest is a_k, so $f(a_k|\boldsymbol{\mu},\mathrm{C},\boldsymbol{\Theta},\underline{a}_{\backslash k})$ can be obtained by considering parameters in Equation (4.22) other than a_k as fixed numbers:

$$
\begin{aligned}
f(a_k \mid \mathbf{X},\boldsymbol{\mu},\mathrm{C},\boldsymbol{\Theta},\underline{a}_{\backslash k}) & \propto a_k^{-(\alpha+\nu_{C0}/2)-1} \times \exp\left(-\frac{\beta+2\mathrm{C}_{0,kk}^{-1}}{a_k}\right) \\
& = \mathrm{IG}\left(\alpha+\frac{\nu_{C0}}{2},\beta+2\mathrm{C}_{0,kk}^{-1}\right)
\end{aligned}
\tag{4.23}
$$

For the fully conditional distribution $p(\nu_0|\boldsymbol{\mu},\mathrm{C},\underline{\mu}_0,\mathrm{C}_0,\boldsymbol{\Sigma}_0,\underline{a})$, the parameter of interest is ν_0, so $p(\nu_0|\boldsymbol{\mu},\mathrm{C},\underline{\mu}_0,\mathrm{C}_0,\boldsymbol{\Sigma}_0,\underline{a})$ can be obtained by considering parameters in Equation (4.22) other than ν_0 as fixed numbers:

$$
p(\nu_0 \mid \mathbf{X},\boldsymbol{\mu},\mathrm{C},\underline{\mu}_0,\mathrm{C}_0,\boldsymbol{\Sigma}_0,\underline{a}) \propto \frac{\left(\prod_{i=1}^{n_s}\left|\mathrm{C}_i\right|^{-\frac{\nu_0}{2}}\right) \times \left|\boldsymbol{\Sigma}_0\right|^{\frac{n_s\nu_0}{2}}}{2^{n_s n\nu_0/2} \cdot \Gamma_n(\nu_0/2)^{n_s}}
\tag{4.24}
$$

When no prior information is available, it is preferable to adopt (roughly) non-informative priors for all hyper-parameters $\boldsymbol{\Theta} = (\underline{\mu}_0,\mathrm{C}_0,\boldsymbol{\Sigma}_0,\nu_0)$. This can be achieved by adopting $\underline{\mu}_{\mu 0}$ = zero vector, $\mathrm{C}_{\mu 0}$ = diagonal matrix with large diagonals (e.g., 10^4), $\nu_{C0} = n+1$, $\alpha = 0.5$, β = a small number (e.g., 10^{-4}), $\lambda_{\Sigma 0} = n+2$, and $\boldsymbol{\Psi}_{\Sigma 0}$ = diagonal matrix with large diagonals (e.g., 10^4). With these settings, ($\boldsymbol{\mu}$,C,$\boldsymbol{\Theta}$,$\underline{a}$) samples ~ f($\boldsymbol{\mu}$,C,$\boldsymbol{\Theta}$,$\underline{a}$|X) can be drawn by the following GS algorithm:

1. Initialize ($\boldsymbol{\mu}$,C,$\boldsymbol{\Theta}$,$\underline{a}$) samples at arbitrary values.
2. (For i = 1, …, n_s): Draw $\underline{\mu}_i$ sample ~ $f(\underline{\mu}_i|\boldsymbol{\mu}_{\backslash i},C,\boldsymbol{\Theta},\underline{a})$:

$$\underline{\mu}_i \sim N\left\{\left(C_0^{-1}+m_iC_i^{-1}\right)^{-1}\left(C_0^{-1}\underline{\mu}_0+C_i^{-1}\sum_{j=1}^{m_i}\underline{X}_i^{(j)}\right),\left(C_0^{-1}+m_iC_i^{-1}\right)^{-1}\right\} \quad (4.25)$$

3. (For i = 1, …, n_s): Draw C_i sample ~ $f(C_i|\boldsymbol{\mu},C_{\backslash i},\boldsymbol{\Theta},\underline{a})$:

$$C_i \sim IW\left\{\boldsymbol{\Sigma}_0+\sum_{j=1}^{m_i}\left(\underline{X}_i^{(j)}-\underline{\mu}_i\right)\left(\underline{X}_i^{(j)}-\underline{\mu}_i\right)^T, m_i+\nu_0\right\} \quad (4.26)$$

4. Draw $\underline{\mu}_0$ sample ~ $f(\underline{\mu}_0|\boldsymbol{\mu},C,C_0,\boldsymbol{\Sigma}_0,\nu_0,\underline{a})$:

$$\underline{\mu}_0 \sim N\left\{\left(C_{\mu_0}^{-1}+n_sC_0^{-1}\right)^{-1}\left(C_{\mu_0}^{-1}\underline{\mu}_{\mu_0}+C_0^{-1}\sum_{i=1}^{n_s}\underline{\mu}_i\right),\left(C_{\mu_0}^{-1}+n_sC_0^{-1}\right)^{-1}\right\} \quad (4.27)$$

5. Draw C_0 sample ~ $f(C_0|\boldsymbol{\mu},C,\underline{\mu}_0,\boldsymbol{\Sigma}_0,\nu_0,\underline{a})$:

$$C_0 \sim IW\left\{\boldsymbol{\Sigma}_{C_0}+\sum_{i=1}^{n_s}\left(\underline{\mu}_i-\underline{\mu}_0\right)\left(\underline{\mu}_i-\underline{\mu}_0\right)^T, n_s+\nu_{C0}\right\} \quad (4.28)$$

6. Draw $\boldsymbol{\Sigma}_0$ sample ~ $f(\boldsymbol{\Sigma}_0|\boldsymbol{\mu},C,\underline{\mu}_0,C_0,\nu_0,\underline{a})$:

$$\boldsymbol{\Sigma}_0 \sim W\left\{\left(\boldsymbol{\Psi}_{\Sigma_0}^{-1}+\sum_{i=1}^{n_s}C_i^{-1}\right)^{-1}, n_s\nu_0+\lambda_{\Sigma_0}\right\} \quad (4.29)$$

7. Draw ν_0 sample from the following discrete distribution:

$$p(\nu_0 \mid \mathbf{X},\boldsymbol{\mu},C,\underline{\mu}_0,C_0,\boldsymbol{\Sigma}_0,\underline{a}) \propto \frac{\prod_{i=1}^{n_s}\left|C_i\right|^{-\frac{\nu_0}{2}}\cdot\left|\boldsymbol{\Sigma}_0\right|^{\frac{n_s\nu_0}{2}}}{2^{n_s n\nu_0/2}\cdot\Gamma_n\left(\nu_0/2\right)^{n_s}} \qquad \nu_0=n,n+1,\ldots,1000$$

(4.30)

8. (For k = 1, ..., n): Draw a_k sample ~ $f(a_k|\boldsymbol{\mu},\mathbf{C},\underline{\mu}_0,\mathbf{C}_0,\boldsymbol{\Sigma}_0,\nu_0,\underline{a}_{\backslash k})$:

$$a_k \sim IG\left(\alpha + \frac{\nu_{C_0}}{2}, \beta + 2C_{0,kk}^{-1}\right) \tag{4.31}$$

9. Cycles 2–8 for T time steps to obtain T samples for $(\boldsymbol{\mu},\mathbf{C},\boldsymbol{\Theta},\underline{a})$.

The samples after the burn-in period (T_b) are collected. These samples are distributed as $f(\boldsymbol{\mu},\mathbf{C},\boldsymbol{\Theta},\underline{a}|\mathbf{X})$. If we only keep the samples of the hyper-parameters $\boldsymbol{\Theta}$, they are distributed as $f(\boldsymbol{\Theta}|\mathbf{X})$. For the GS algorithm, samples at nearby time steps are correlated. To reduce sample correlation and also to reduce subsequent computation, samples are collected at time steps T_b+1, T_b+Δt+1, ..., T. There are in total N = $(T\text{-}T_b)/\Delta t$ samples, and let us denote the resulting $\boldsymbol{\Theta}$ samples by $\{\boldsymbol{\Theta}_k$: k = 1, ..., N}. They have been "trained" by the soil database **X** about the site-uniqueness characteristics.

4.1.3.3 Behavior of the prior model constructed by the hyper-parameter samples

The hyper-parameter samples $\{\boldsymbol{\Theta}_k$: k = 1, ..., N} can be adopted to construct a prior model for a future target site (e.g., the Taipei site). To illustrate this prior model, the samples $\{\boldsymbol{\Theta}_k$: k = 1, ..., N} can simulate the site-specific mean vector and covariance matrix $(\underline{\mu}^{(h)}, \mathbf{C}^{(h)})$ for a "hypothetical site" by the following steps:

1. Randomly draw $\boldsymbol{\Theta}$ from $\{\boldsymbol{\Theta}_k$: k = 1, ..., N} and denote these samples by $\left(\hat{\underline{\mu}}_0, \hat{C}_0, \hat{\Sigma}_0, \hat{\nu}_0\right)$.
2. Draw $\underline{\mu}^{(h)} \sim N\left(\hat{\underline{\mu}}_0, \hat{C}_0\right)$ and $\mathbf{C}^{(h)} \sim IW\left(\hat{\Sigma}_0, \hat{\nu}_0\right)$.
3. Cycle Steps 1 and 2 to obtain the desirable number of $(\underline{\mu}^{(h)}, \mathbf{C}^{(h)})$ samples.

The $(\mu_1^{(h)}, \sigma_1^{(h)}, \mu_2^{(h)}, \sigma_2^{(h)}, \delta_{12}^{(h)})$ sample values can be extracted from the $(\underline{\mu}^{(h)}, \mathbf{C}^{(h)})$ samples. The light grey cross markers and light grey histogram in Figure 4.11 show the $(\mu_1^{(h)}, \sigma_1^{(h)}, \mu_2^{(h)}, \sigma_2^{(h)}, \delta_{12}^{(h)})$ sample values for 100 such hypothetical sites. These samples illustrate the prior model constructed by the hyper-parameter samples $\{\boldsymbol{\Theta}_k$: k = 1, ..., N}. The dark circles and dark histogram in Figure 4.11 show the actual statistics of the n_s = 72 sites in the database. It is clear that the hypothetical site statistics are similar to the actual site statistics in the database, suggesting that the hyper-parameter samples $\{\boldsymbol{\Theta}_k$: k = 1, ..., N} have effectively learned the site-uniqueness characteristics in the database. Another way to illustrate the similarity is shown in Figure 4.9b. Each $(\underline{\mu}^{(h)}, \mathbf{C}^{(h)})$ sample can be represented as an ellipse in the (X_1, X_2) space, and this ellipse can be mapped back to the (Y_1, Y_2) space

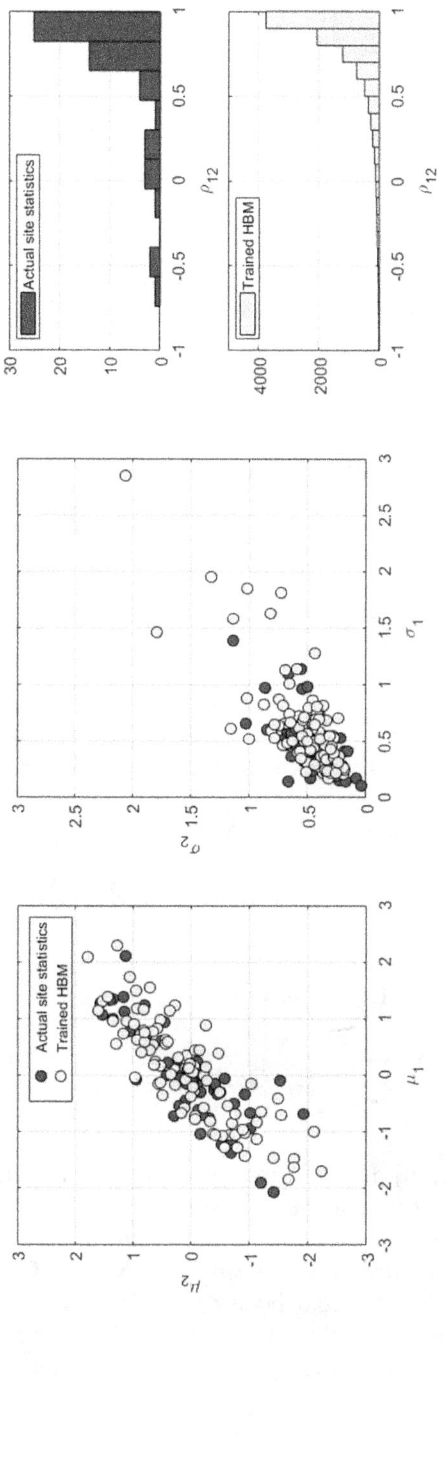

Figure 4.11 Behavior of the hyper-parameter samples $\{\boldsymbol{\Theta}_k: k = 1, \ldots, N\}$.

through the inverse-Johnson transform (the ellipse may be skewed by the nonlinear transform). Figure 4.9b shows the skewed ellipses that represent the 100 hypothetical sites in the (Y_1, Y_2) space. Figure 4.9b can be compared to the database data in Figure 4.9a.

4.1.3.4 Bayesian parameter estimation for the Taipei site

The purpose here is to estimate the underlying site-specific mean vector ($\underline{\mu}_s$) and site-specific covariance matrix (C_s) for the Taipei site. The assumed model is **M** = the HBM. From the perspective of the Taipei site, $f(\underline{\mu}_s,C_s|\mathbf{X})$ is the prior model that is trained by the database **X**, and this prior model is characterized by the hyper-parameter samples $\{\mathbf{\Theta}_k$: k = 1, …, N$\}$. Let us assume the Taipei site is governed by the same hyper-parameters (as seen in Figure 4.10) as the database sites. Given the i.i.d. Taipei-site data $\mathbf{X}_s = \{\underline{X}_s^{(1)}, \underline{X}_s^{(2)}, \ldots, \underline{X}_s^{(ms)}\}$ (m_s = 9) in Table 4.2, the purpose is to further update this prior model $f(\underline{\mu}_s,C_s|\mathbf{X})$ into the posterior model $f(\underline{\mu}_s,C_s|\mathbf{X},\mathbf{X}_s)$ by further conditioning on the Taipei-site data $\mathbf{X}_s$. This posterior model is quasi-site-specific because it is based on the prior model trained by **X** and then further updated by $\mathbf{X}_s$. According to the Total Probability Theorem,

$$f(\underline{\mu}_s,C_s \mid \mathbf{X},\mathbf{X}_s) = \int f(\underline{\mu}_s,C_s \mid \mathbf{\Theta},\mathbf{X},\mathbf{X}_s)f(\mathbf{\Theta} \mid \mathbf{X},\mathbf{X}_s)d\mathbf{\Theta} \tag{4.32}$$

Note that given $\mathbf{\Theta}$, **X** and $(\underline{\mu}_s,C_s)$ become independent, hence, $f(\underline{\mu}_s,C_s|\mathbf{\Theta},\mathbf{X},\mathbf{X}_s) = f(\underline{\mu}_s,C_s|\mathbf{\Theta},\mathbf{X}_s)$. Let us further assume $f(\mathbf{\Theta}|\mathbf{X},\mathbf{X}_s) \approx f(\mathbf{\Theta}|\mathbf{X})$. This assumption is reasonable because **X** contains 72 sites, whereas $\mathbf{X}_s$ only contains one site. The information in $\{\mathbf{X},\mathbf{X}_s\}$ is roughly the same as that in **X** (73 sites vs. 72 sites). As a result, we have

$$f(\underline{\mu}_s,C_s \mid \mathbf{X},\mathbf{X}_s) \approx \int f(\underline{\mu}_s,C_s \mid \mathbf{\Theta},\mathbf{X}_s)f(\mathbf{\Theta} \mid \mathbf{X})d\mathbf{\Theta} \tag{4.33}$$

According to the Law of Large Numbers,

$$f(\underline{\mu}_s,C_s \mid \mathbf{X},\mathbf{X}_s) \approx \frac{1}{N}\sum_{k=1}^{N} f(\underline{\mu}_s,C_s \mid \mathbf{\Theta}_k,\mathbf{X}_s) \tag{4.34}$$

where $\{\mathbf{\Theta}_k$: k = 1, …, N$\}$ are the samples drawn from $f(\mathbf{\Theta}|\mathbf{X})$. The samples $(\underline{\mu}_s,C_s) \sim f(\underline{\mu}_s,C_s|\mathbf{\Theta},\mathbf{X}_s)$ can also be drawn using a sub-GS algorithm. In this sub-GS algorithm, let us divide the variables into $\underline{\mu}_s$ and C_s, and their samples can be obtained by drawing $\underline{\mu}_s \sim f(\underline{\mu}_s|C_s,\mathbf{\Theta},\mathbf{X}_s)$ and $C_s \sim f(C_s|\underline{\mu}_s,\mathbf{\Theta},\mathbf{X}_s)$ in an alternating manner. Note that both $f(\underline{\mu}_s|C_s,\mathbf{\Theta},\mathbf{X}_s)$ and $f(C_s|\underline{\mu}_s,\mathbf{\Theta},\mathbf{X}_s)$ have analytical expressions (see Table 2.1). $f(\underline{\mu}_s|C_s,\mathbf{\Theta},\mathbf{X}_s)$ can be categorized as the third scenario in Table 2.1, where $\underline{\mu}_s$ is the unknown parameter, $\mathbf{X}_s = \{\underline{X}_s^{(1)},$

$\underline{X}_s^{(2)}, \ldots, \underline{X}_s^{(ms)}\}$ are the observations, $f(\mathbf{X}_s|\underline{\mu}_s,\mathbf{C}_s,\mathbf{\Theta})$ is the likelihood (which is multivariate normal with known $\mathbf{C}_s$), and $f(\underline{\mu}_s|\mathbf{C}_s,\mathbf{\Theta}) = f(\underline{\mu}_s|\mathbf{\Theta})$ is the prior PDF (which is multivariate normal). It is clear from the third scenario in Table 2.1 that the posterior PDF $f(\underline{\mu}_s|\mathbf{C}_s,\mathbf{\Theta},\mathbf{X}_s)$ is the following multivariate normal PDF:

$$f(\underline{\mu}_s \mid \mathbf{C}_s,\mathbf{\Theta},\mathbf{X}_s) = \mathrm{N}\left\{\left(\mathbf{C}_0^{-1} + m_s\mathbf{C}_s^{-1}\right)^{-1}\left(\mathbf{C}_0^{-1}\underline{\mu}_0 + \sum_{j=1}^{m_s}\mathbf{C}_s^{-1}\underline{X}_s^{(j)}\right),\left(\mathbf{C}_0^{-1} + m_s\mathbf{C}_s^{-1}\right)^{-1}\right\} \tag{4.35}$$

where $\underline{\mu}_0$ and $\mathbf{C}_0$ are given because they are within $\mathbf{\Theta}$. $f(\mathbf{C}_s|\underline{\mu}_s,\mathbf{\Theta},\mathbf{X}_s)$ can be categorized as the fourth scenario in Table 2.1:

$$f(\mathbf{C}_s \mid \underline{\mu}_s,\mathbf{\Theta},\mathbf{X}_s) = \mathrm{IW}\left\{\mathbf{\Sigma}_0 + \sum_{j=1}^{m_s}\left(\underline{X}_s^{(j)} - \underline{\mu}_s\right)\left(\underline{X}_s^{(j)} - \underline{\mu}_s\right)^T, m_s + \nu_0\right\} \tag{4.36}$$

As a result, the samples $(\underline{\mu}_s,\mathbf{C}_s) \sim f(\underline{\mu}_s,\mathbf{C}_s|\mathbf{X},\mathbf{X}_s)$ can be obtained by the following procedure:

1. Randomly draw $\mathbf{\Theta}$ from $\{\mathbf{\Theta}_k$: k = 1, …, N$\}$ and denote these samples by $\hat{\mathbf{\Theta}} = \left(\hat{\underline{\mu}}_0, \hat{\mathbf{C}}_0, \hat{\mathbf{\Sigma}}_0, \hat{\nu}_0\right)$.
2. Conduct the sub-GS sampler to draw $(\underline{\mu}_s,\mathbf{C}_s)$ samples $\sim f(\underline{\mu}_s,\mathbf{C}_s \mid \hat{\mathbf{\Theta}},\mathbf{X}_s)$:

$$\underline{\mu}_s \sim \mathrm{N}\left\{\left(\hat{\mathbf{C}}_0^{-1} + m_s\mathbf{C}_s^{-1}\right)^{-1}\left(\hat{\mathbf{C}}_0^{-1}\hat{\underline{\mu}}_0 + \sum_{j=1}^{m_s}\mathbf{C}_s^{-1}\underline{X}_s^{(j)}\right),\left(\hat{\mathbf{C}}_0^{-1} + m_s\mathbf{C}_s^{-1}\right)^{-1}\right\} \tag{4.37}$$

   a. Initialize $(\underline{\mu}_s,\mathbf{C}_s)$ samples at arbitrary values.
   b. Update $\underline{\mu}_s$ sample:
   c. Update $\mathbf{C}_s$ sample:

$$\mathbf{C}_s \sim \mathrm{IW}\left\{\hat{\mathbf{\Sigma}}_0 + \sum_{j=1}^{m_s}\left(\underline{X}_s^{(j)} - \underline{\mu}_s\right)\left(\underline{X}_s^{(j)} - \underline{\mu}_s\right)^T, m_s + \hat{\nu}_0\right\} \tag{4.38}$$

   d. Cycle Steps b and c for T′ times to obtain T′ samples for $(\underline{\mu}_s,\mathbf{C}_s)$. There is also a burn-in period, denoted by T'_b. We simply take $T' = T'_b + 1$ such that each hyper-parameter sample $\hat{\mathbf{\Theta}}$ will produce one $(\underline{\mu}_s,\mathbf{C}_s)$ sample.
3. Cycle Steps 1 and 2 to obtain the desirable number of $(\underline{\mu}_s,\mathbf{C}_s)$ samples.

These $(\underline{\mu}_s,\mathbf{C}_s)$ samples are distributed as $f(\underline{\mu}_s,\mathbf{C}_s|\mathbf{X},\mathbf{X}_s)$, the posterior (quasi-site-specific) model. Figure 4.12 demonstrates the samples of $(\underline{\mu}_s,\mathbf{C}_s)$ in terms

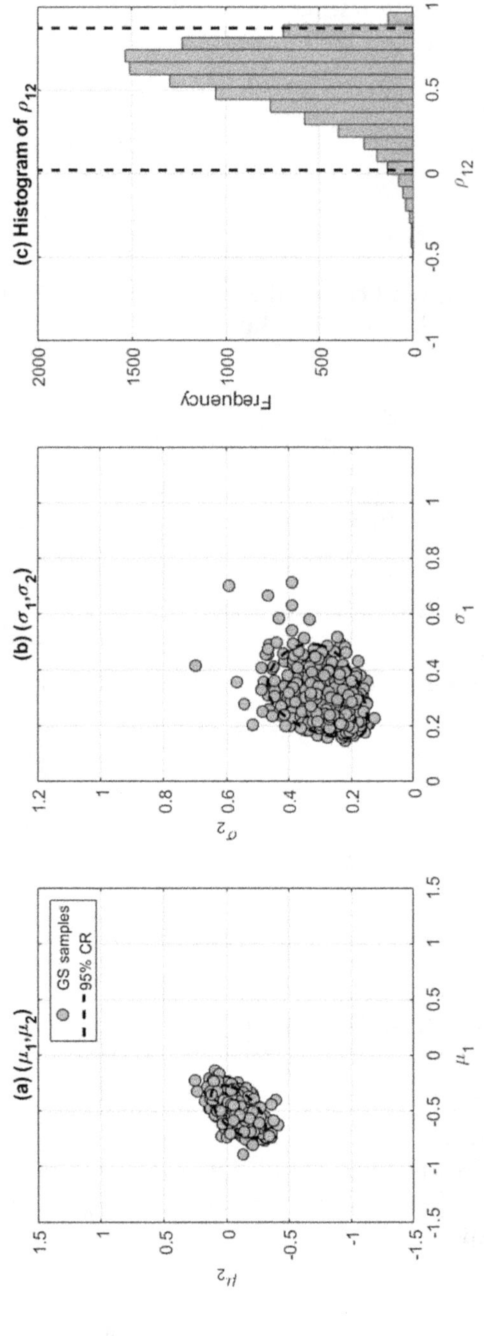

Figure 4.12 Posterior samples of ($\underline{\mu}_s$, $\mathbf{C}_s$) (quasi-site-specific model).

of (μ_1, σ_1, μ_2, σ_2, ρ_{12}). Figure 4.12 can be compared to Figure 4.7. The main difference between Figure 4.7 [which is based on f(μ_s,C_s|X_s), the site-specific model] and Figure 4.12 [which is based on f(μ_s,C_s|X,X_s), the quasi-site-specific model] is that the former adopts an non-informative prior model, whereas the latter adopts an informative prior model constructed by the HBM based on **X**. Figures 4.7a and 4.12a show that the (μ_1, μ_2) samples have similar centers, suggesting that the site-specific trend and quasi-site-specific trend are similar. Figures 4.7b and 4.12b show that the (σ_1, σ_2) samples for the quasi-site-specific model are smaller than those for the site-specific model, suggesting that the quasi-site-specific model has smaller transformation uncertainty. More importantly, the scatter of the quasi-site-specific (μ_s,C_s) samples in Figure 4.12 is less than the scatter of the site-specific (μ_s,C_s) samples in Figure 4.7. This is because the quasi-site-specific model adopts the informative prior model trained by **X**.

4.1.3.5 Bayesian prediction for the Taipei site

Now consider the depths at the Taipei site that are different from the nine depths. Recall that the q_t values at these depths are known (as shown in Figure 4.13a),

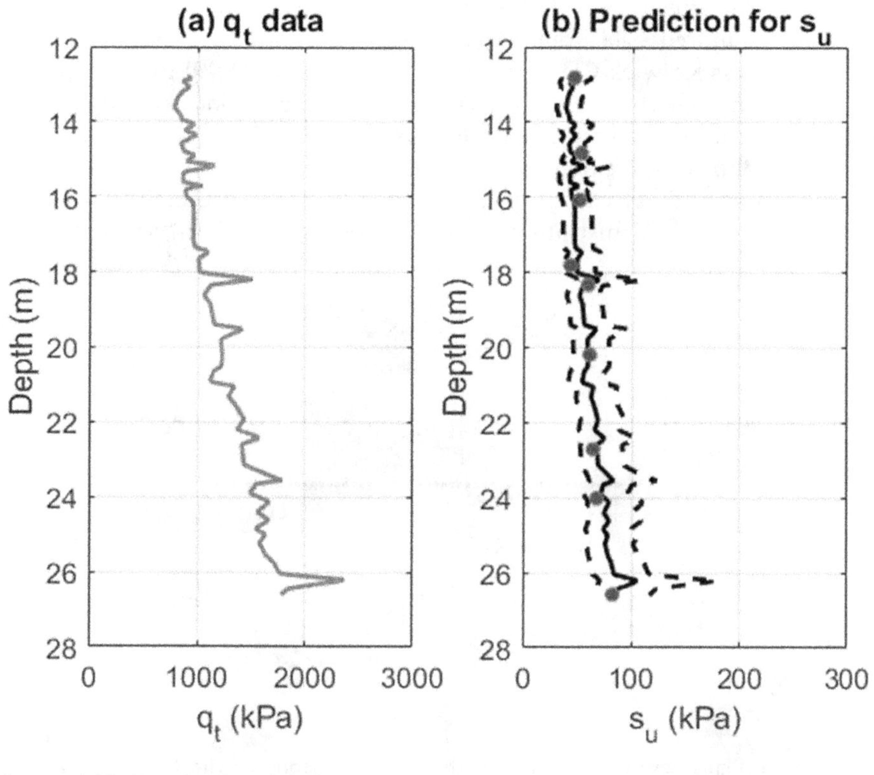

Figure 4.13 Results for the clay layer at a Taipei site (quasi-site-specific model).

and it is desirable to predict the s_u/σ'_v values at these depths. The same procedure for Bayesian prediction in the previous section can be adopted, but now the (μ_s,C_s) samples in Figure 4.12 for the quasi-site-specific model are adopted to obtain the posterior median and 95% CI profiles of s_u/σ'_v. The resulting posterior median and 95% CI profiles of s_u/σ'_v are shown in Figure 4.13b. The width of the 95% CI is much narrower than that for the site-specific model (Figure 4.8b), yet all validating data (the dots in Figure 4.13b) still lie with the 95% CI. It is also much narrower than that for the generic model (Figure 4.6b). As a result, the quasi-site-specific model in the current section has much less transformation uncertainty compared to the generic model and to the site-specific model.

4.2 SPATIALLY VARIABLE DATA

The CPT is popular in geotechnical site investigation because it can provide a nearly continuous vertical profile of the cone tip resistance (q_t). The q_t profile can be used to back calculate some features of the soil spatial variability such as spatial trend, variance, and auto-correlation structure. Now consider a test site at South Parklands, Adelaide (South Australia) with more than two hundred CPT soundings performed in a stiff, overconsolidated clay known as Keswick Clay (Jaksa 1995; Jaksa et al. 1999) to the depth of about 5 m. Among the soundings, 51 soundings were conducted along a line on the ground surface with a horizontal interval of 0.5 m. The 51 soundings spanning a horizontal extent (x direction) of 25 m, from x = 30 m, 30.5 m, …, 54.5 m, and 55 m (see Figure 4.14). The q_t data for all soundings between z = 1.5 m and z = 5 m (sampling depth interval = 0.02 m) are shown in

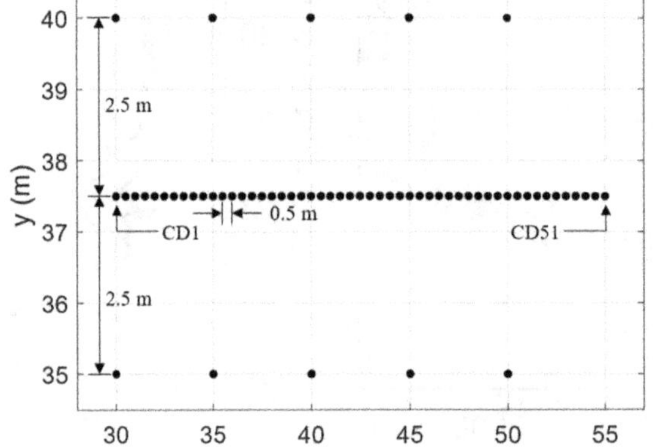

Figure 4.14 Plan view of a subset of the CPT locations at the South Parklands site.

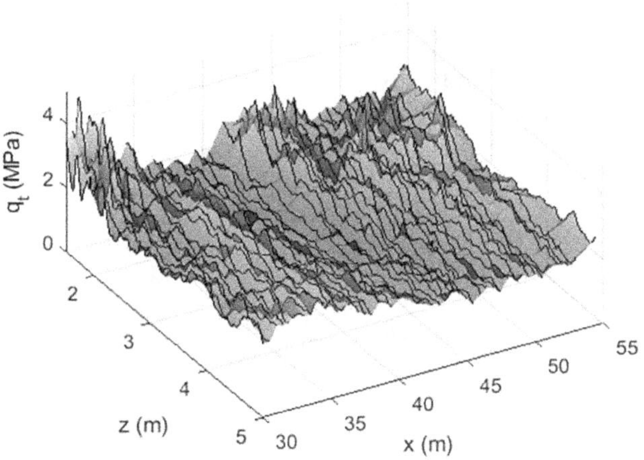

Figure 4.15 q_t data for all CPTs between z = 1.5 m and z = 5 m.

Figure 4.15. The depth range from 1.5 to 5 m is considered as Keswick Clay by Jaksa (1995) and Jaksa et al. (1999).

4.2.1 One-dimensional (1D) data

To demonstrate 1D data analysis, the q_t data for the CPT at x = 30 m are first analyzed (the data are shown in Figure 4.16). The q_t data is modeled

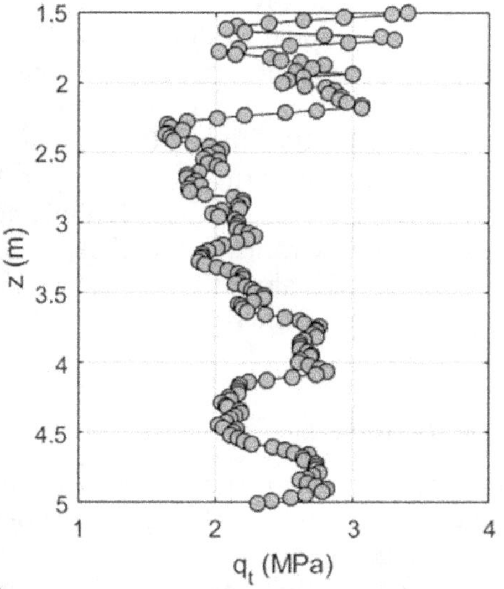

Figure 4.16 q_t data for the CPT at x = 30 m.

as $\xi(z) = t(z) + w(z)$, where $\xi(z) = q_t(z)$, $t(z)$ is the depth trend, and $w(z)$ is the spatial variability. The spatial variability $w(z)$ is modeled as zero-mean stationary normal random field with the Whittle–Matérn (WM) auto-correlation model (see Section 2.6):

$$\rho(\Delta z) = \frac{2}{\Gamma(v)} \cdot \left(\frac{\sqrt{\pi} \cdot \Gamma(v+0.5) \cdot |\Delta z|}{\Gamma(v) \cdot \delta} \right)^{v} K_v \left(\frac{2\sqrt{\pi} \cdot \Gamma(v+0.5) \cdot |\Delta z|}{\Gamma(v) \cdot \delta} \right) \quad (4.39)$$

The trend function $t(z)$ is modeled as the quadratic function of depth:

$$t(z) = a + b \cdot z + c \cdot z^2 \quad (4.40)$$

where $\{a, b, c\}$ are unknown trend parameters, denoted by $\underline{\theta}^{(t)}$.

4.2.1.1 Bayesian parameter estimation

Given the data $\underline{D} = [\xi(z_1)\ \xi(z_2)\ \ldots\ \xi(z_m)]$, the purpose of parameter estimation is to estimate the unknown parameters including the auto-covariance parameters for the spatial variability $\underline{w}$, denoted by $\underline{\theta}^{(w)} = \{\sigma, \delta, v\}$, where σ is the standard deviation for $\underline{w}$, and δ and v are the scale of fluctuation and smoothness, respectively, and the trend parameters $\underline{\theta}^{(t)} = \{a, b, c\}$. Let $\underline{\theta} = \{\underline{\theta}^{(w)}, \underline{\theta}^{(t)}\}$ denote the collection of all unknown parameters. Because there are no analytical Bayesian solutions for the posterior PDF $f(\underline{\theta}|\underline{D})$ (no conjugate priors), the transitional Markov chain Monte Carlo (TMCMC) algorithm (see Section 3.4.1) is adopted to draw posterior samples $\underline{\theta} \sim f(\underline{\theta}|\underline{D})$. The TMCMC algorithm requires the calculation of the likelihood function $f(\underline{D}|\underline{\theta})$, which is exemplified by Example 3.13. The prior PDFs are such that $\ln(\sigma)$ is uniform over ln(0.01 MPa) to ln(10 MPa), $\ln(\delta)$ is uniform over ln(0.01 m) to ln(1 m), $\ln(v)$ is uniform over ln(0.1) to ln(3), a is uniform over 0 to 10 MPa, b is uniform over −5 to 5 MPa/m, and c is uniform over −1 to 1 MPa/m^2. The sample size for each TMCMC stage is taken to be N = 2000. In the end of the TMCMC algorithm, 2000 samples of $\underline{\theta} \sim f(\underline{\theta}|\underline{D})$ are obtained. Figure 4.17 shows how the posterior samples of $\underline{\theta}$.

4.2.1.2 Bayesian prediction

For Bayesian prediction, the same procedure described in Example 3.13 can be adopted to simulate a new profile of q_t data (denoted by $\underline{D}'$) that follows the same trend, variance, and auto-correlation parameters. Figure 4.18 shows two such simulation examples as dots (q_t) and dashed lines (trend). In the same figure, the q_t profile at x = 30 m (i.e., the training data $\underline{D}$) is also

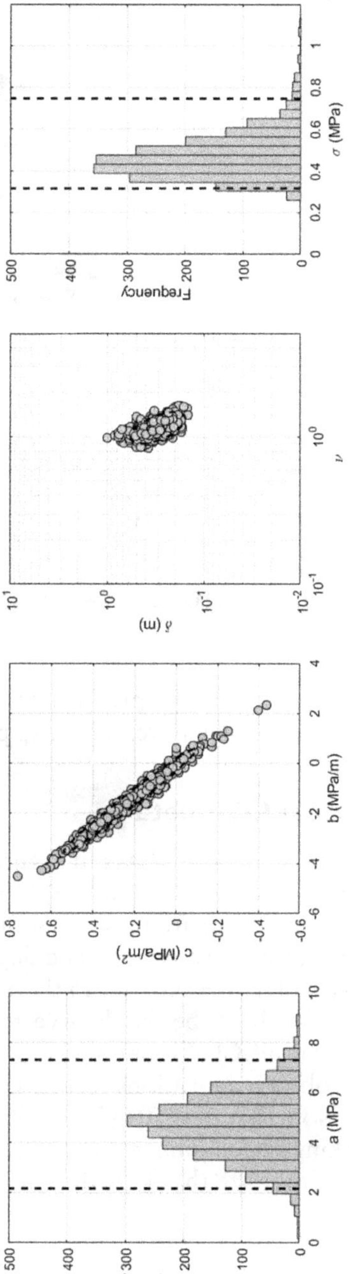

Figure 4.17 Posterior samples of $\underline{\theta}$ (1D spatial data).

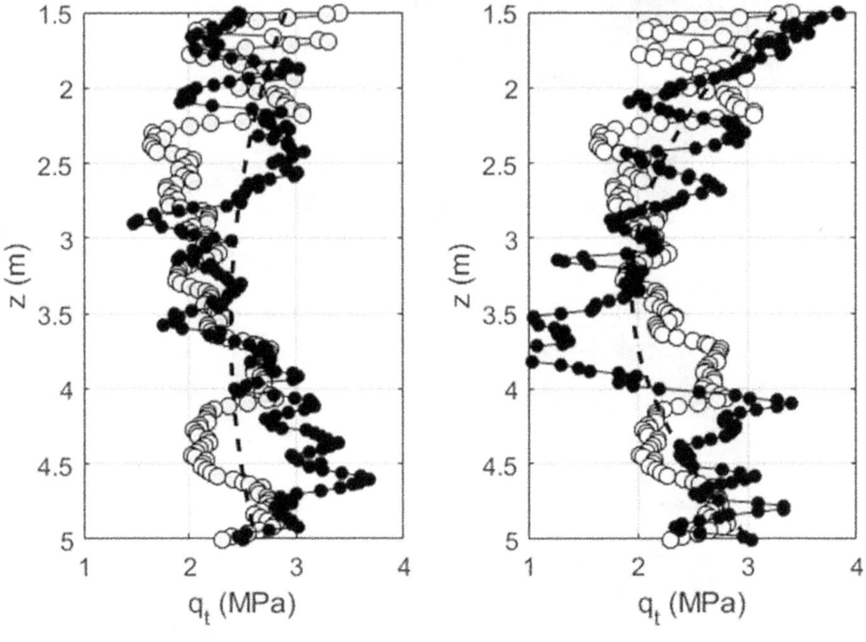

Figure 4.18 Two sets of simulated $\underline{D}'$ (1D spatial data).

shown for comparison. The simulated $\underline{D}'$ profiles exhibit similar characteristics (trend, variance, SOF, smoothness, etc.) to the q_t profile at x = 30 m.

4.2.2 Three-dimensional (3D) data

Now consider the data from all 51 soundings in Figure 4.15. Let m_h be the number of all CPT soundings. The sounding at x = 42 m is intentionally left out of the training data; hence, there are $m_h = 50$ soundings. Let m_z be the number of depths observed at each sounding. The depths are from z = 1.5 to 5 m with sampling depth interval = 0.02 m, so there are $m_z = 126$ depths. Let $\underline{\xi}^{(i)} = (\xi_1^{(i)}, \xi_2^{(i)}, \ldots, \xi_{mz}^{(i)})^T \in \mathbf{R}^{mz\times 1}$ be the data vector observed at the ith sounding. Let us assume that all CPT soundings measure the same depths $(z_1, z_2, \ldots, z_{mz})$, that is, the observed locations form a lattice in 3D. There are in total $m = m_h \times m_z$ data. Note that $m = 50\times 126 = 6300$ is very large. Let $\underline{D} \in \mathbf{R}^{m\times 1}$ be formed by stacking the column vectors $\underline{\xi}^{(1)}, \underline{\xi}^{(2)}, \ldots$, and $\underline{\xi}^{(mh)}$ vertically. It is customary to hypothesize that $\underline{D}$ is the summation of the trend $\underline{t}$ and the spatial variability $\underline{w}$:

$$\underline{D} = \underline{t} + \underline{w} \tag{4.41}$$

where $\underline{t} \in \mathbf{R}^{m\times 1}$ contains the trend function values evaluated at the m locations, and $\underline{w} \in \mathbf{R}^{m\times 1}$ contains the spatial variability values at the m locations. The auto-covariance matrix for $\underline{w}$ is denoted by $\mathbf{C} \in \mathbf{R}^{m\times m}$:

$$C = \sigma^2 \times \begin{bmatrix} 1 & \rho(x_1 - x_2, y_1 - y_2, z_1 - z_2) & \rho(x_1 - x_3, y_1 - y_3, z_1 - z_3) & \cdots & \rho(x_1 - x_m, y_1 - y_m, z_1 - z_m) \\ & 1 & \rho(x_2 - x_3, y_2 - y_3, z_2 - z_3) & \cdots & \rho(x_2 - x_m, y_2 - y_m, z_2 - z_m) \\ & & 1 & & \\ & \text{SYM.} & & \ddots & \rho(x_{m-1} - x_m, y_{m-1} - y_m, z_{m-1} - z_m) \\ & & & & 1 \end{bmatrix} \tag{4.42}$$

where ρ is the auto-correlation function for $\underline{w}$. Here, a common "separability" assumption is adopted: the auto-correlation model is separable in the horizontal and vertical directions. Namely,

$$\rho(\Delta x, \Delta y, \Delta z) = \rho_h(\Delta x, \Delta y) \times \rho_z(\Delta z) \tag{4.43}$$

where ρ_h and ρ_z are the horizontal and vertical auto-correlation functions for $\underline{w}$, respectively. Both are modeled by the WM auto-correlation model:

$$\rho_h(\Delta x, \Delta y) = \frac{2}{\Gamma(v_h)} \cdot \left(\frac{\sqrt{\pi} \cdot \Gamma(v_h + 0.5) \cdot \sqrt{\Delta x^2 + \Delta y^2}}{\Gamma(v_h) \cdot \delta_h} \right)^{v_h} K_{v_h}\left(\frac{2\sqrt{\pi} \cdot \Gamma(v_h + 0.5) \cdot \sqrt{\Delta x^2 + \Delta y^2}}{\Gamma(v_h) \cdot \delta_h} \right) \tag{4.44}$$

$$\rho_z(\Delta z) = \frac{2}{\Gamma(v_z)} \cdot \left(\frac{\sqrt{\pi} \cdot \Gamma(v_z + 0.5) \cdot |\Delta z|}{\Gamma(v_z) \cdot \delta_z} \right)^{v_z} K_{v_z}\left(\frac{2\sqrt{\pi} \cdot \Gamma(v_z + 0.5) \cdot |\Delta z|}{\Gamma(v_z) \cdot \delta_z} \right) \tag{4.45}$$

where δ_h and δ_z are the horizontal and vertical SOFs, respectively; v_h and v_z are the horizontal and vertical smoothnesses, respectively. The unknown auto-covariance parameters include $\underline{\theta}^{(w)} = \{\sigma, \delta_h, \delta_z, v_h, v_z\}$.

For the data in Figure 4.15, let us consider the following trend function t(x,y,z):

$$t(x, y, z) = a + b \cdot z + c \cdot z^2 + d \cdot x + e \cdot x^2 + f \cdot x \times z \tag{4.46}$$

where $\underline{\theta}^{(t)} = \{a, b, c, d, e, f\}$ are unknown trend parameters. Namely, the trend is a quadratic function of (x,z). It is not a function of y because the soundings lie in a constant-y plane. The log-likelihood function is

$$\ln\left[f(\underline{D} \mid \underline{\theta})\right] = -\frac{m}{2} \times \ln(2\pi) - \frac{1}{2}\ln(|\mathbf{C}|) - 0.5 \times (\underline{D} - \underline{t})^{T} \mathbf{C}^{-1} (\underline{D} - \underline{t}) \quad (4.47)$$

The evaluation of the above likelihood function is time consuming because it may require the matrix inversion $\mathbf{C}^{-1}$ and the matrix determinant $|\mathbf{C}|$, and the size of $\mathbf{C} \in \mathbf{R}^{m\times m}$ is huge (i.e., 6300 × 6300). Moreover, the Bayesian analysis requires repetitive evaluations of the likelihood function, so it may become computationally prohibitive for 3D problems. Nevertheless, with the separability assumption, we show in the following that the computational cost can be greatly reduced by adopting the Kronecker-product expressions. To begin with, first note that with the separability assumption, the $\mathbf{C}$ matrix can be expressed as the following Kronecker product:

$$\mathbf{C} = \sigma^2 \times (\mathbf{R}_h \otimes \mathbf{R}_z) \quad (4.48)$$

where $\mathbf{A} \otimes \mathbf{B}$ denotes the Kronecker product between two matrices $\mathbf{A}$ and $\mathbf{B}$:

$$\mathbf{A} \otimes \mathbf{B} = \begin{bmatrix} a_{11}\mathbf{B} & a_{12}\mathbf{B} & \cdots & a_{1n}\mathbf{B} \\ a_{21}\mathbf{B} & a_{22}\mathbf{B} & \cdots & a_{2n}\mathbf{B} \\ \vdots & \vdots & \ddots & \vdots \\ a_{m1}\mathbf{B} & a_{m2}\mathbf{B} & \cdots & a_{mn}\mathbf{B} \end{bmatrix} \quad (4.49)$$

$\mathbf{R}_h \in \mathbf{R}^{mh\times mh}$ is the horizontal auto-correlation matrix among the m_h sounding locations:

$$\mathbf{R}_h = \begin{bmatrix} 1 & \rho_h(x_1 - x_2, y_1 - y_2) & \rho_h(x_1 - x_3, y_1 - y_3) & \cdots & \rho_h(x_1 - x_{mh}, y_1 - y_{mh}) \\ & 1 & \rho_h(x_2 - x_3, y_2 - y_3) & \cdots & \rho_h(x_2 - x_{mh}, y_2 - y_{mh}) \\ & & 1 & & \\ & & & \ddots & \rho_h(x_{mh-1} - x_{mh}, y_{mh-1} - y_{mh}) \\ \text{SYM.} & & & & 1 \end{bmatrix} \quad (4.50)$$

$\mathbf{R}_z \in \mathbf{R}^{mz\times mz}$ is the vertical auto-correlation matrix among the m_z sampling depths:

$$\mathbf{R}_z = \begin{bmatrix} 1 & \rho_z(z_1 - z_2) & \rho_z(z_1 - z_3) & \cdots & \rho_z(z_1 - z_{mz}) \\ & 1 & \rho_z(z_2 - z_3) & \cdots & \rho_z(z_2 - z_{mz}) \\ & & 1 & & \\ & & & \ddots & \rho_z(z_{mz-1} - z_{mz}) \\ \text{SYM.} & & & & 1 \end{bmatrix} \quad (4.51)$$

The size of **C** is m×m, which is huge. However, the sizes of $\mathbf{R}_h$ and $\mathbf{R}_z$ are only $m_h \times m_h$ and $m_z \times m_z$, respectively. With the following derivations, the likelihood function can be calculated efficiently:

$$\begin{aligned}
\ln\left[f(\underline{D} \mid \underline{\theta})\right] &= -\frac{m}{2}\ln(2\pi) - \frac{1}{2}\ln\left(\left|\sigma^2(\mathbf{R}_h \otimes \mathbf{R}_z)\right|\right) \\
&\quad - 0.5(\underline{D}-\underline{t})^T\left[\sigma^2(\mathbf{R}_h \otimes \mathbf{R}_z)\right]^{-1}(\underline{D}-\underline{t}) \\
&= -\frac{m}{2}\ln(2\pi) - m\ln(\sigma) - \frac{1}{2}\ln\left(\left|\mathbf{R}_h \otimes \mathbf{R}_z\right|\right) - \frac{1}{2\sigma^2}(\underline{D}-\underline{t})^T\left[(\mathbf{R}_h \otimes \mathbf{R}_z)\right]^{-1}(\underline{D}-\underline{t}) \\
&= -\frac{m}{2}\ln(2\pi) - m\ln(\sigma) - \frac{1}{2}\ln\left(\left|\mathbf{R}_h\right|^{m_z}\left|\mathbf{R}_z\right|^{m_h}\right) - \frac{1}{2\sigma^2}(\underline{D}-\underline{t})^T\left(\mathbf{R}_h^{-1} \otimes \mathbf{R}_z^{-1}\right)(\underline{D}-\underline{t}) \\
&= -\frac{m}{2}\ln(2\pi) - m\ln(\sigma) - \ln\left(\left|\mathbf{L}_h\right|^{m_z}\left|\mathbf{L}_z\right|^{m_h}\right) \\
&\quad - \frac{1}{2\sigma^2}(\underline{D}-\underline{t})^T\left(\left(\mathbf{L}_h^{-T}\mathbf{L}_h^{-1}\right) \otimes \left(\mathbf{L}_z^{-T}\mathbf{L}_z^{-1}\right)\right)(\underline{D}-\underline{t}) \\
&= -\frac{m}{2}\ln(2\pi) - m\ln(\sigma) - m_z\sum_{i=1}^{m_h}\ln(L_{h,ii}) - m_h\sum_{i=1}^{m_z}\ln(L_{z,ii}) \\
&\quad - \frac{1}{2\sigma^2}(\underline{D}-\underline{t})^T\left(\left(\mathbf{L}_h^{-1} \otimes \mathbf{L}_z^{-1}\right)^T\left(\mathbf{L}_h^{-1} \otimes \mathbf{L}_z^{-1}\right)\right)(\underline{D}-\underline{t}) \\
&= -\frac{m}{2}\ln(2\pi) - m\ln(\sigma) - m_z\sum_{i=1}^{m_h}\ln(L_{h,ii}) - m_h\sum_{i=1}^{m_z}\ln(L_{z,ii}) \\
&\quad - \frac{1}{2\sigma^2}\left[\left(\mathbf{L}_h^{-1} \otimes \mathbf{L}_z^{-1}\right)(\underline{D}-\underline{t})\right]^T\left[\left(\mathbf{L}_h^{-1} \otimes \mathbf{L}_z^{-1}\right)(\underline{D}-\underline{t})\right] \\
&= -\frac{m}{2}\ln(2\pi) - m\ln(\sigma) - m_z\sum_{i=1}^{m_h}\ln(L_{h,ii}) - m_h\sum_{i=1}^{m_z}\ln(L_{z,ii}) \\
&\quad - \frac{1}{2\sigma^2}\mathrm{vec}\left[\mathbf{L}_z^{-1}\mathrm{mat}(\underline{D}-\underline{t})\mathbf{L}_h^{-T}\right]^T\mathrm{vec}\left[\mathbf{L}_z^{-1}\mathrm{mat}(\underline{D}-\underline{t})\mathbf{L}_h^{-T}\right]
\end{aligned} \tag{4.52}$$

where $\mathbf{L}_h \in \mathbf{R}^{mh\times mh}$ and $\mathbf{L}_z \in \mathbf{R}^{mz\times mz}$ are lower triangular Cholesky decompositions for $\mathbf{R}_h$ and $\mathbf{R}_z$, respectively; $\mathrm{mat}(\underline{D}\text{-}\underline{t}) \in \mathbf{R}^{mz\times mh}$ is reshaped from $(\underline{D}\text{-}\underline{t}) \in \mathbf{R}^{m\times 1}$; $\mathrm{vec}(\mathbf{A}_{mz\times mh}) \in \mathbf{R}^{m\times 1}$ is formed by stacking the columns in **A** vertically. The above derivations have adopted the following identities: $(\mathbf{A}\otimes\mathbf{B})^{-1} = \mathbf{A}^{-1}\otimes\mathbf{B}^{-1}$, $(\mathbf{A}\otimes\mathbf{B})^T = \mathbf{A}^T\otimes\mathbf{B}^T$, $(\mathbf{AB})\otimes(\mathbf{CD}) = (\mathbf{A}\otimes\mathbf{C})(\mathbf{B}\otimes\mathbf{D})$, $(\mathbf{A}^T\otimes\mathbf{B})\times\mathrm{vec}(\mathbf{C}) = \mathrm{vec}(\mathbf{BCA})$, and $|\mathbf{A}_{m\times m}\otimes\mathbf{B}_{n\times n}| = |\mathbf{A}|^n|\mathbf{B}|^m$. One can see that now the evaluation of the likelihood function does not require the matrix inversion $\mathbf{C}^{-1}$ and the matrix determinant $|\mathbf{C}|$.

4.2.2.1 Bayesian parameter estimation

Given the data $\underline{D}$ from all soundings, the purpose of parameter estimation is to estimate the unknown parameters $\underline{\theta} = \{\underline{\theta}^{(w)}, \underline{\theta}^{(t)}\}$. Again, the TMCMC algorithm is adopted to draw posterior samples $\underline{\theta} \sim f(\underline{\theta}|\underline{D})$. The prior PDFs are similar to those for the 1D analysis: $\ln(\sigma)$ is uniform over ln(0.01 MPa) to ln(10 MPa), $\ln(\delta_z)$ is uniform over ln(0.01 m) to ln(1 m), $\ln(\nu_z)$ is uniform over ln(0.1) to ln(3), a is uniform over 0 to 10 MPa, b is uniform over −5 to 5 MPa/m, and c is uniform over −1 to 1 MPa/m^2. The extra 3D parameters have the following priors: $\ln(\delta_h)$ is uniform over ln(0.1 m) to ln(10 m), $\ln(\nu_h)$ is uniform over ln(0.1) to ln(3), d is uniform over −5 to 5 MPa/m, e is uniform over −1 to 1 MPa/m^2, and f is uniform over −1 to 1 MPa/m^2. The sample size for each TMCMC stage is taken to be N = 2000. In the end of the TMCMC algorithm, 2000 samples of $\underline{\theta} \sim f(\underline{\theta}|\underline{D})$ are obtained. Figure 4.19 shows how the posterior samples of $\underline{\theta}$ (d, e, and f are not shown for brevity). Recall that TMCMC can estimate the evidence $f(\underline{D})$ as a by-product. The log-evidence estimate for is 1056.2.

Figure 4.20 shows the surface plot of one sample for the trend function, whereas Figure 4.21 shows the projections of the trend at soundings with x = 30, 35, 40, 45, 50, and 55. Although the trend function in general follows the overall large-scale trend of the data (Figure 4.20), it fails to provide a satisfactory fit to the small-scale trends for most soundings (Figure 4.21). A more complicated trend model (a quadratic function is not complicated enough) is necessary to fit these small-scale trends, but it is challenging to determine the proper algebraic form.

4.2.2.2 *Bayesian prediction (conditional random field simulation)*

Now consider the simulation of the random field at m_h' new horizontal (x,y) locations, and, at each (x,y) location, the random field is simulated at the same m_z depths. Let $\underline{\xi}'^{(1)}$, $\underline{\xi}'^{(2)}$, …, and $\underline{\xi}'^{(mh')} \in \mathbf{R}^{mz\times1}$ denote the data vector at these simulation locations. There are in total $m' = m_h' \times m_z$ data points to simulate. Let $\underline{D}' \in \mathbf{R}^{m'\times1}$ be formed by stacking the column vectors $\underline{\xi}'^{(1)}$, $\underline{\xi}'^{(2)}$, …, and $\underline{\xi}^{(mh')}$ vertically. Let $\underline{t}' \in \mathbf{R}^{m'\times1}$ be the corresponding trend vector. The purpose of Bayesian prediction is to simulate conditional random field samples at these new locations, that is, to draw $\underline{D}'$ samples $\sim f(\underline{D}'|\underline{D})$.

If $\underline{\theta} = \{\underline{\theta}^{(w)},\underline{\theta}^{(t)}\}$ are known, it can be shown that $\underline{D}$ and $\underline{D}'$ are jointly normal with the following mean vectors and covariance matrices:

$$E\left(\begin{bmatrix}\underline{D}'\\ \underline{D}\end{bmatrix}\middle|\underline{\theta}\right)=\begin{bmatrix}\underline{t}'\\ \underline{t}\end{bmatrix}\qquad COV\left(\begin{bmatrix}\underline{D}'\\ \underline{D}\end{bmatrix}\middle|\underline{\theta}\right)=\begin{bmatrix}C_{\underline{D}'} & C_{\underline{D}'\underline{D}}\\ C_{\underline{D}'\underline{D}}^{T} & C_{\underline{D}}\end{bmatrix} \tag{4.53}$$

where E(.) denotes a mean vector; COV(.) denotes a covariance matrix; $C_{\underline{D}}$ = covariance matrix for $\underline{D}$; $C_{\underline{D}'}$ = covariance matrix for $\underline{D}'$; $C_{\underline{D}'\underline{D}}$ = covariance matrix between $\underline{D}'$ and $\underline{D}$. Due to the separability of the auto-correlation function, $C_{\underline{D}} = \sigma^2\times(\mathbf{R}_h\otimes\mathbf{R}_z)$, $C_{\underline{D}'} = \sigma^2 \times (\mathbf{R}_{h'}\otimes\mathbf{R}_z)$, and $C_{\underline{D}'\underline{D}} = \sigma^2\times(\mathbf{R}_{h'h}\otimes\mathbf{R}_z)$; $\mathbf{R}_{h'} \in \mathbf{R}^{mh'\times mh'}$ is the horizontal auto-correlation matrix among the m'_h simulation locations; $\mathbf{R}_{h'h} \in \mathbf{R}^{mh'\times mh}$ is the horizontal cross-correlation matrix

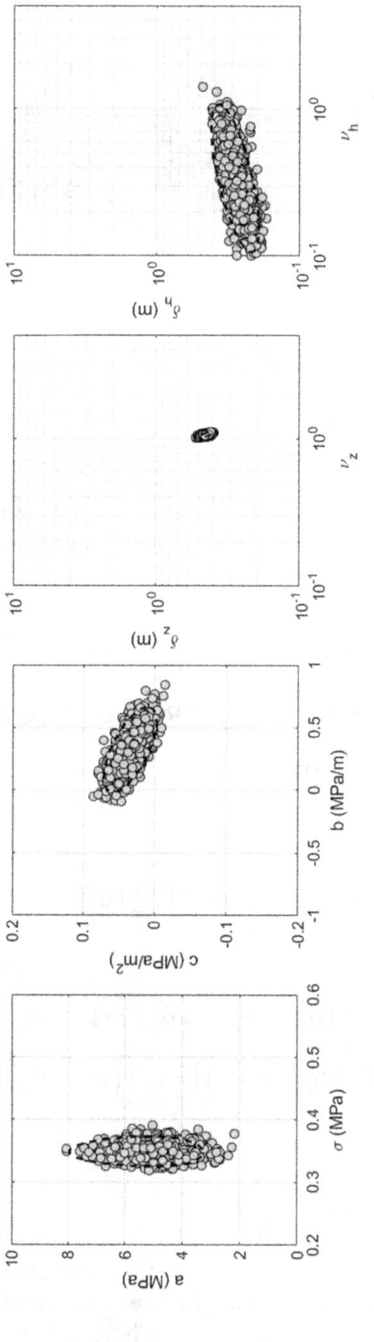

Figure 4.19 Posterior samples of $\underline{\theta}$ (3D spatial data; quadratic trend model).

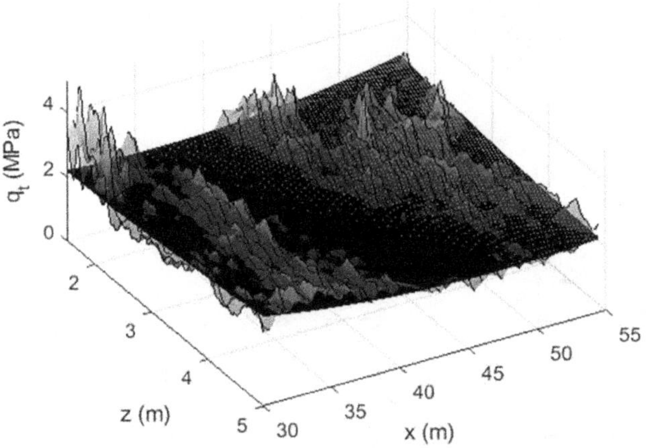

Figure 4.20 Surface plot of one sample for the trend function (quadratic trend model).

between the m'_h simulation locations and m_h sounding locations. Because $\underline{D}$ and $\underline{D}'$ are jointly normal, given $\underline{D}$, $\underline{D}'$ is still multivariate normal with the following mean vector and covariance matrix:

$$\begin{aligned}
&E\left(\underline{D}' \mid \underline{\theta},\underline{D}\right) = E\left(\underline{D}' \mid \underline{\theta}\right) + COV\left(\underline{D}',\underline{D} \mid \underline{\theta}\right) COV\left(\underline{D} \mid \underline{\theta}\right)^{-1}\left[\underline{D} - E\left(\underline{D} \mid \underline{\theta}\right)\right] \\
&= \underline{t}' + \mathbf{C}_{\underline{D}'\underline{D}}\mathbf{C}_{\underline{D}}^{-1}\left(\underline{D} - \underline{t}\right) = \underline{t}' + \left[\sigma^2\left(\mathbf{R}_{h'h} \otimes \mathbf{R}_z\right)\right]\left[\sigma^2\left(\mathbf{R}_h \otimes \mathbf{R}_z\right)\right]^{-1}\left(\underline{D} - \underline{t}\right) \\
&= \underline{t}' + \left[\left(\mathbf{R}_{h'h}\mathbf{R}_h^{-1}\right) \otimes \mathbf{I}_{mz\times mz}\right] vec\left[mat\left(\underline{D} - \underline{t}\right)\right] = \underline{t}' + vec\left[mat\left(\underline{D} - \underline{t}\right)\mathbf{R}_h^{-1}\mathbf{R}_{h'h}^{T}\right]
\end{aligned} \tag{4.54}$$

$$\begin{aligned}
&COV\left(\underline{D}' \mid \underline{\theta},\underline{D}\right) = COV\left(\underline{D}' \mid \underline{\theta}\right) - COV\left(\underline{D}',\underline{D} \mid \underline{\theta}\right) COV\left(\underline{D} \mid \underline{\theta}\right)^{-1} \\
&COV\left(\underline{D}',\underline{D} \mid \underline{\theta}\right)^{T} \\
&= \mathbf{C}_{\underline{D}'} - \mathbf{C}_{\underline{D}'\underline{D}}\mathbf{C}_{\underline{D}}^{-1}\mathbf{C}_{\underline{D}'\underline{D}}^{T} = \sigma^2\left[\left(\mathbf{R}_{h'} \otimes \mathbf{R}_z\right) - \left(\mathbf{R}_{h'h} \otimes \mathbf{R}_z\right)\left(\mathbf{R}_h \otimes \mathbf{R}_z\right)^{-1}\left(\mathbf{R}_{h'h} \otimes \mathbf{R}_z\right)^{T}\right] \\
&= \sigma^2\left[\mathbf{R}_{h'} \otimes \mathbf{R}_z - \left(\left[\mathbf{R}_{h'h}\mathbf{R}_h^{-1}\mathbf{R}_{h'h}^{T}\right] \otimes \mathbf{R}_z\right)\right] = \sigma^2\left[\left(\mathbf{R}_{h'} - \mathbf{R}_{h'h}\mathbf{R}_h^{-1}\mathbf{R}_{h'h}^{T}\right) \otimes \mathbf{R}_z\right] \\
&= \sigma^2\left(\mathbf{R}_{h'|h} \otimes \mathbf{R}_z\right)
\end{aligned} \tag{4.55}$$

where $\mathbf{R}_{h'|h}$ is defined as $\mathbf{R}_{h'} - \mathbf{R}_{h'h}\mathbf{R}_{h}^{-1}\mathbf{R}_{h'h}^{T}$. A conditional sample of $\underline{D}' \sim f(\underline{D}'|\underline{\theta},\underline{D})$ can be readily simulated if the Cholesky decomposition of $COV(\underline{D}'|\underline{\theta},\underline{D})$ is obtained. Note that the lower triangular Cholesky decomposition of $\mathbf{R}_{h'|h}\otimes\mathbf{R}_z$ is equal to $\mathbf{L}_{h'|h}\otimes\mathbf{L}_z$, where $\mathbf{L}_{h'|h}$ is the lower triangular

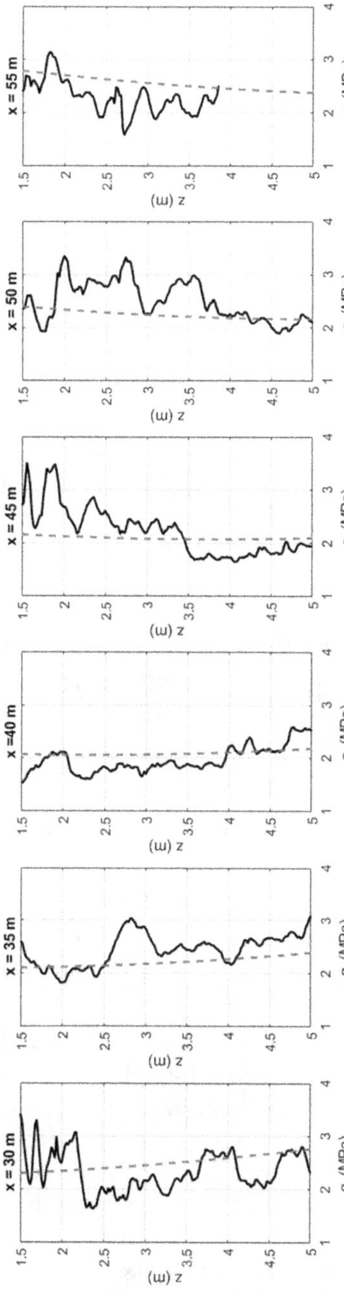

Figure 4.21 Projections of the trend at soundings with x = 30, 35, 40, 45, 50, and 55.

Cholesky decomposition for $\mathbf{R}_{h'|h}$. As a result, the conditional sample of $\underline{D}'$ ~ $f(\underline{D}'|\underline{\theta},\underline{D})$ can be obtained efficiently by the following equation:

$$\begin{aligned}\underline{D}' &= \underline{t}' + \mathrm{vec}\left[\mathrm{mat}\left(\underline{D}-\underline{t}\right)\mathbf{R}_h^{-1}\mathbf{R}_{h'h}^T\right] + \sigma\left(\mathbf{L}_{h'|h}\otimes\mathbf{L}_z\right)\underline{Z} \\ &= \underline{t}' + \mathrm{vec}\left[\mathrm{mat}\left(\underline{D}-\underline{t}\right)\mathbf{R}_h^{-1}\mathbf{R}_{h'h}^T\right] + \sigma\left(\mathbf{L}_{h'|h}\otimes\mathbf{L}_z\right)\mathrm{vec}\left[\mathrm{mat}\left(\mathbf{Z}\right)\right] \\ &= \underline{t}' + \mathrm{vec}\left[\mathrm{mat}\left(\underline{D}-\underline{t}\right)\mathbf{R}_h^{-1}\mathbf{R}_{h'h}^T\right] + \sigma\times\mathrm{vec}\left[\mathbf{L}_z\mathrm{mat}\left(\mathbf{Z}\right)\mathbf{L}_{h'|h}^T\right]\end{aligned} \tag{4.56}$$

However, $\underline{\theta} = \{\underline{\theta}^{(w)},\underline{\theta}^{(t)}\}$ are actually unknown. The purpose of the conditional random field simulation is to draw $\underline{D}'$ samples ~ $f(\underline{D}'|\underline{D})$. According to the Total Probability Theorem and Law of Large Numbers:

$$f(\underline{D}'\mid\underline{D}) = \int f(\underline{D}'\mid\underline{\theta},\underline{D})f(\underline{\theta}\mid\underline{D})d\underline{\theta} \approx \frac{1}{N}\sum_{k=1}^{N} f(\underline{D}'\mid\underline{\theta}_k,\underline{D}) \tag{4.57}$$

where $\underline{\theta}_k = \{a_k, b_k, c_k, \sigma_k, \delta_{h,k}, \delta_{z,k}, v_{h,k}, v_{z,k}\}$ is the kth TMCMC sample drawn from $f(\underline{\theta}|\underline{D})$.

As a result, the $\underline{D}'$ samples ~ $f(\underline{D}'|\underline{D})$ can be obtained using the following procedure:

1. Randomly draw $\underline{\theta}$ from the TMCMC samples $\{\underline{\theta}_k$: k = 1, …, N$\}$ and denote this sample by $\left(\hat{a},\hat{b},\hat{c},\hat{d},\hat{e},\hat{f},\hat{\sigma},\hat{\delta}_h,\hat{\delta}_z,\hat{v}_h,\hat{v}_z\right)$.
2. Based on $\left(\hat{a},\hat{b},\hat{c},\hat{d},\hat{e},\hat{f}\right)$, $\hat{\underline{t}}$ and $\hat{\underline{t}}'$ can be obtained. Based on $\left(\hat{\delta}_h,\hat{\delta}_z,\hat{v}_h,\hat{v}_z\right)$, $(\hat{\mathbf{R}}_h,\hat{\mathbf{R}}_{h'h},\hat{\mathbf{R}}_{h'|h},\hat{\mathbf{R}}_z)$ can be computed, and so are $(\hat{\mathbf{L}}_{h'|h},\hat{\mathbf{L}}_z)$.
3. Draw a sample $\underline{D}'$ using the following equation:

$$\underline{D}' = \hat{\underline{t}}' + \mathrm{vec}\left[\mathrm{mat}\left(\underline{D}-\hat{\underline{t}}\right)\hat{\mathbf{R}}_h^{-1}\hat{\mathbf{R}}_{h'h}^T\right] + \hat{\sigma}\times\mathrm{vec}\left[\hat{\mathbf{L}}_z\mathrm{mat}\left(\underline{Z}\right)\hat{\mathbf{L}}_{h'|h}^T\right] \tag{4.58}$$

4. Cycle Steps 1–3 to obtain the desirable number of $\underline{D}'$ sample.

Figure 4.22 shows the conditional random field simulation results at x = 42 m (recall that there is a sounding here, but the q_t data is intentionally left out of the training data $\underline{D}$), x = 40 m (there is a sounding at 40 m), x = 40.01 m (right next to 40 m), and 40.1 m (close to 40 m). The light dots represent one conditional random field sample, the solid line represents the conditional median, and the dashed lines represent the conditional 95% CI. The figure also shows the relevant q_t data for comparison. The dark dots in Figure 4.22a show the actual q_t data at x = 42 m. Note that the actual q_t data are intentionally left out of the training data $\underline{D}$, so the comparison in Figure 4.22a (between the simulation results and the actual q_t data) is a genuine validation. The conditional random field median (solid line) does

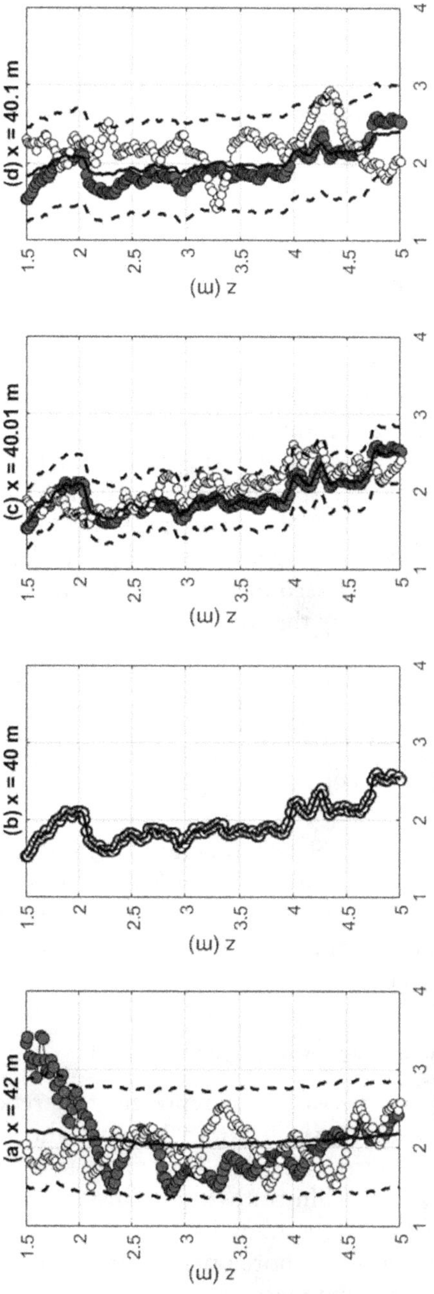

Figure 4.22 Conditional random field simulation results at several locations (quadratic trend model).

not follow the actual data well, possibly due to the fact that the adopted quadratic trend function model is not complicated enough. The dark dots in Figure 4.22b,c,d show the actual q_t data at x = 40 m. The perfect match between the simulation results and the actual q_t data in Figure 4.22b is correct because the actual q_t data at x = 40 m is within the training data $\underline{D}$. It is remarkable that the 95% CI grows quickly from zero width in Figure 4.22b to noticeable width in Figure 4.22c even though the distance from x = 40 to 40.01 m is only 0.01 m.

4.2.3 Three-dimensional (3D) data – GPR method for trend modeling

Given the complicated trend of the data, it is challenging to determine the proper algebraic form for the trend function. The GPR method adopts a different strategy of modeling the trend without explicitly writing down the algebraic form for the trend function (Yoshida et al. 2021; Ching et al. 2023). The GPR method adopts a zero-mean stationary normal random field as the prior model for the spatial trend function. Note that this zero-mean is the "prior mean" for trend, not the posterior mean. The posterior mean for the trend will follow the general trend of the data $\underline{D}$. The spatial trend t(x,y,z) is modeled as a zero-mean stationary normal random field with auto-correlation defined by the squared exponential model (QExp):

$$\rho^{(t)}\left(x_i - x_j, y_i - y_j, z_i - z_j\right) = \rho_h^{(t)}\left(x_i - x_j, y_i - y_j\right)\rho_z^{(t)}\left(z_i - z_j\right)$$
$$= \exp\left(-\pi\frac{\left(x_i - x_j\right)^2}{\delta_h^{(t)2}} - \pi\frac{\left(y_i - y_j\right)^2}{\delta_h^{(t)2}}\right)\exp\left(-\pi\frac{\left(z_i - z_j\right)^2}{\delta_z^{(t)2}}\right) \tag{4.59}$$

where $\rho^{(t)}$ denotes the auto-correlation function for the t(x,y,z); $\rho^{(t)}{}_h$ and $\rho^{(t)}{}_z$ denote the horizontal and vertical auto-correlations, respectively; $\delta^{(t)}{}_h$ and $\delta^{(t)}{}_z$ are the SOFs of t(x,y,z) in the horizontal and vertical directions, respectively. The QExp model is adopted for trend because t(x,y,z) is assumed to be smooth. The standard deviation for t(x,y,z) is denoted by $\sigma^{(t)}$.

4.2.3.1 *Karhunen–Loève expansion for the trend*

For the GPR method, it is necessary to represent the trend function as a linear combination of a few basis vectors whose weights are independent zero-mean normal random variables (i.e., Karhunen–Loève expansion, e.g., Ghanem and Spanos 1991). To illustrate this expansion, consider the "union lattice" that consists of the m_h sounding locations and m'_h simulation locations. In total, there are $m_h + m_{h'}$ horizontal locations in this union lattice (the subscript "u" denotes "union lattice"), and there are $m_u = m + m'$ data points ($m = m_h \times m_z$ and $m' = m_{h'} \times m_z$). Let the trend on this union lattice be denoted by $\underline{t}_u = \{\underline{t}, \underline{t}'\} \in \mathbf{R}^{(m+m')\times 1}$, and the auto-covariance matrix for $\underline{t}_u$

is denoted by $\mathbf{C}^{(t)}_{u} \in \mathbf{R}^{(m+m')\times(m+m')}$. Note that the QExp auto-correlation for the trend is separable in the horizontal and vertical directions (see Equation 4.59). Therefore,

$$\mathbf{C}_{u}^{(t)} = \sigma^{(t)2} \times \left(\mathbf{R}_{h,u}^{(t)} \otimes \mathbf{R}_{z}^{(t)}\right) \tag{4.60}$$

where $\sigma^{(t)}$ denotes the standard deviation for the trend; $\mathbf{R}^{(t)}_{h,u} \in \mathbf{R}^{(mh+mh')\times(mh+mh')}$ is the horizontal QExp auto-correlation matrix for the trend among the $m_h + m_{h'}$ horizontal locations; $\mathbf{R}^{(t)}_{z} \in \mathbf{R}^{mz\times mz}$ is the vertical QExp auto-correlation matrix for the trend among the m_z depths. Let us denote the ith eigenvalue and eigenvector of $\mathbf{R}^{(t)}_{h,u}$ by $(\lambda_{h,i}, \varphi_{h,i})$. Let the eigenvalues and eigenvectors are ranked such that $\lambda_{h,1} > \lambda_{h,2} > \lambda_{h,3} \ldots$ For computational efficiency, let us only consider the first p eigenvalues such that they consist of at least 99% of the total sum: $(\lambda_{h,1} + \lambda_{h,2} + \ldots + \lambda_{h,p}) \geq 0.99\times(\lambda_{h,1} + \lambda_{h,2} + \ldots + \lambda_{h,(mh+mh')})$. The value of p is usually much smaller than $m_h + m_{h'}$. With the first p eigenvalues and eigenvectors, $\mathbf{R}^{(t)}_{h,u}$ have the following approximate expression:

$$\mathbf{R}_{h,u}^{(t)} \approx \mathbf{\Phi}_{h,u} \times \mathbf{\Omega}_{h,u} \times \mathbf{\Phi}_{h,u}^{T} \tag{4.61}$$

where $\mathbf{\Omega}_{h,u} \in \mathbf{R}^{p\times p}$ is a diagonal matrix that contains the first p eigenvalues $(\lambda_{h,1}, \lambda_{h,2} \ldots \lambda_{h,p})$; $\mathbf{\Phi}_{h,u} \in \mathbf{R}^{(mh+mh')\times p}$ contains the first p eigenvectors of $\mathbf{R}^{(t)}_{h,u}$. Here, $\mathbf{\Phi}_{h,u}$ can be expressed as the vertical stack of two block matrices:

$$\mathbf{\Phi}_{h,u} = \begin{bmatrix} \mathbf{\Phi}_{h} \\ \mathbf{\Phi}_{h'} \end{bmatrix} \tag{4.62}$$

where $\mathbf{\Phi}_{h} \in \mathbf{R}^{mh\times p}$ contains the eigenvectors at the m_h sounding locations; $\mathbf{\Phi}_{h'} \in \mathbf{R}^{mh'\times p}$ contains the eigenvectors at the $m_{h'}$ simulation locations. A similar expansion can be done for $\mathbf{R}^{(t)}_{z}$:

$$\mathbf{R}_{z}^{(t)} \approx \mathbf{\Phi}_{z} \times \mathbf{\Omega}_{z} \times \mathbf{\Phi}_{z}^{T} \tag{4.63}$$

where $\mathbf{\Omega}_{z} \in \mathbf{R}^{q\times q}$ is a diagonal matrix that contains the first q eigenvalues (the first q eigenvalues constitute at least 99% of the total sum); $\mathbf{\Phi}_{z} \in \mathbf{R}^{mz\times q}$ contains the first q eigenvectors. The value of q is also usually much smaller than m_z. With the expressions in Equations (4.61) and (4.63), $\mathbf{C}^{(t)}_{u}$ has the following expression:

$$\mathbf{C}_{u}^{(t)} = \sigma^{(t)2} \times \left(\mathbf{\Phi}_{h,u} \times \mathbf{\Omega}_{h,u} \times \mathbf{\Phi}_{h,u}^{T}\right) \otimes \left(\mathbf{\Phi}_{z} \times \mathbf{\Omega}_{z} \times \mathbf{\Phi}_{z}^{T}\right) = \begin{bmatrix} \mathbf{\Phi}_{h} \otimes \mathbf{\Phi}_{z} \\ \mathbf{\Phi}_{h'} \otimes \mathbf{\Phi}_{z} \end{bmatrix} \times \left[\sigma^{(t)2}\left(\mathbf{\Omega}_{h,u} \otimes \mathbf{\Omega}_{z}\right)\right] \times \begin{bmatrix} \mathbf{\Phi}_{h} \otimes \mathbf{\Phi}_{z} \\ \mathbf{\Phi}_{h'} \otimes \mathbf{\Phi}_{z} \end{bmatrix}^{T} \tag{4.64}$$

As a result, the trend vector $\underline{t}_u$ can be expressed as

$$\underline{t}_u = \begin{bmatrix} \underline{t} \\ \underline{t}' \end{bmatrix} = \begin{bmatrix} \mathbf{\Phi}_h \otimes \mathbf{\Phi}_z \\ \mathbf{\Phi}_{h'} \otimes \mathbf{\Phi}_z \end{bmatrix} \times \underline{\omega} \tag{4.65}$$

where $\underline{\omega} \in \mathbf{R}^{pq\times1}$ is a zero-mean normal vector with covariance matrix = $\mathbf{\Omega}$ = $\sigma^{(t)2}\times(\mathbf{\Omega}_{h,u}\otimes\mathbf{\Omega}_z)$. Therefore, both $\underline{t}$ and $\underline{t}'$ can be expressed as linear combinations of basis vectors whose weights are independent zero-mean normal random variables:

$$\underline{t} = \left(\mathbf{\Phi}_h \otimes \mathbf{\Phi}_z\right) \times \underline{\omega} \qquad \underline{t}' = \left(\mathbf{\Phi}_{h'} \otimes \mathbf{\Phi}_z\right) \times \underline{\omega} \tag{4.66}$$

Moreover, the matrices containing the basis vectors have Kronecker-product expressions such as $\mathbf{\Phi}_h\otimes\mathbf{\Phi}_z$ and $\mathbf{\Phi}_h'\otimes\mathbf{\Phi}_z$. With this expression, the auto-covariance matrix for $\underline{t}$ is

$$\mathbf{C}^{(t)} = \left(\mathbf{\Phi}_h \otimes \mathbf{\Phi}_z\right) \times \mathbf{\Omega} \times \left(\mathbf{\Phi}_h \otimes \mathbf{\Phi}_z\right)^T \tag{4.67}$$

4.2.3.2 Evaluation of the likelihood function

For the GPR method, $\underline{D} = \underline{t} + \underline{w}$, so $\underline{D}$ is now a zero-mean stationary normal random field with auto-covariance matrix = $\mathbf{C}^{(t)} + \mathbf{C}$, where $\mathbf{C}^{(t)}$ is the auto-covariance matrix for $\underline{t}$ (Equation 4.67); $\mathbf{C}$ is the auto-covariance matrix for $\underline{w}$. For the spatial variability part, its auto-correlation is assumed to follow the WM model with the separability assumption, so

$$\mathbf{C} = \sigma^2 \times \left(\mathbf{R}_h \otimes \mathbf{R}_z\right) \tag{4.68}$$

where σ denotes the standard deviation for the spatial variability; $\mathbf{R}_h \in \mathbf{R}^{mh\times mh}$ is the horizontal WM auto-correlation matrix for the spatial variability; $\mathbf{R}_z \in \mathbf{R}^{mz\times mz}$ is the vertical WM auto-correlation matrix for the spatial variability. Let $\underline{\theta}^{(t)} = \{\sigma^{(t)}, \delta^{(t)}{}_h, \delta^{(t)}{}_z\}$ denote the collection of all auto-covariance parameters for the trend and $\underline{\theta}^{(w)} = \{\sigma, \delta_h, \delta_z, \nu_h, \nu_z\}$ denote the collection of all auto-covariance parameters for the spatial variability (the superscript "(w)" denotes spatial variability), and let $\underline{\theta} = \{\underline{\theta}^{(w)},\underline{\theta}^{(t)}\}$ denote the collection of all unknown parameters. For the GPR method, the log-likelihood function is therefore

$$\ln\left[f(\underline{D} \mid \underline{\theta})\right] = -\frac{m}{2} \times \ln(2\pi) - \frac{1}{2}\ln\left(\left|\mathbf{C}^{(t)} + \mathbf{C}\right|\right) - 0.5 \times \underline{D}^T\left(\mathbf{C}^{(t)} + \mathbf{C}\right)^{-1}\underline{D} \tag{4.69}$$

Note that $\mathbf{C}^{(t)} + \mathbf{C}$ does not have a Kronecker-product expression. As a result, efficient evaluation for the likelihood function such as Equation (4.52) is not applicable. The evaluation of the likelihood function is again time consuming because it may require the matrix inversion and the matrix determinant of a potentially huge matrix $\mathbf{C}^{(t)} + \mathbf{C}$. Although Equation (4.52) is no longer applicable, Appendix 4.1 (also see Ching et al. 2020) shows that the likelihood function can be evaluated by the following formula if $\mathbf{C}^{(t)}$ has the Kronecker-product expression in Equation (4.67):

$$\ln[f(\underline{D} \mid \underline{\theta})] = -\frac{m}{2}\ln(2\pi) - m_z\sum_{i=1}^{m_h}\ln(L_{h,ii}) - m_h\sum_{i=1}^{m_z}\ln(L_{z,ii}) - (m - p\times q)\times\ln(\sigma) - \sum_{i=1}^{n}\ln(L_{ii}) - \frac{1}{2\sigma^2}\left(\underline{\alpha}^T\underline{\alpha} - \underline{\beta}^T\underline{\beta}\right) \quad (4.70)$$

where $(\mathbf{L}_h,\mathbf{L}_z)$ are lower triangular Cholesky decompositions for $(\mathbf{R}_h,\mathbf{R}_z)$;

$$\underline{\alpha} = \text{vec}\left[\mathbf{L}_z^{-1}\text{mat}(\underline{D})\mathbf{L}_h^{-T}\right] \qquad \underline{\beta} = \mathbf{L}^{-1}\times\mathbf{\Omega}^{1/2}\times\text{vec}\left[\mathbf{\Phi}_z^T\mathbf{R}_z^{-1}\text{mat}(\underline{D})\mathbf{R}_h^{-1}\mathbf{\Phi}_h\right] \quad (4.71)$$

$\mathbf{L}$ is the following lower triangular Cholesky decomposition:

$$\mathbf{L}\mathbf{L}^T = \sigma^2\mathbf{I} + \mathbf{\Omega}^{1/2}\times\left[\left(\mathbf{\Phi}_h^T\mathbf{R}_h^{-1}\mathbf{\Phi}_h\right)\otimes\left(\mathbf{\Phi}_z^T\mathbf{R}_z^{-1}\mathbf{\Phi}_z\right)\right]\times\mathbf{\Omega}^{1/2} \quad (4.72)$$

$\mathbf{I}$ is an identity matrix; $(L_{ii}, L_{h,ii}, L_{z,ii})$ are the ith diagonal terms of the lower triangular matrices $(\mathbf{L}, \mathbf{L}_h, \mathbf{L}_z)$. These formulas do not require the time-consuming calculation for large matrices.

4.2.3.3 Bayesian parameter estimation

Given the data $\underline{D}$ from all soundings, the purpose of parameter estimation is to estimate the unknown parameters for the GPR method including the trend parameters $\underline{\theta}^{(t)} = (\sigma^{(t)2}, \delta^{(t)}_h, \delta^{(t)}_z)$ and $\underline{\theta}^{(w)} = (\sigma, \delta_h, \delta_z, \nu_h, \nu_z)$. The prior PDFs are similar to those for the previous example: $\ln(\sigma)$ is uniform over ln(0.01 MPa) to ln(10 MPa), $\ln(\delta_z)$ is uniform over ln(0.01 m) to ln(1 m), $\ln(\nu_z)$ is uniform over ln(0.1) to ln(3), $\ln(\delta_h)$ is uniform over ln(0.1 m) to ln(10 m), and $\ln(\nu_h)$ is uniform over ln(0.1) to ln(3). The extra GPR parameters have the following priors: $\ln(\sigma^{(t)})$ is uniform over ln(0.1 MPa) to ln(10 MPa), $\ln(\delta^{(t)}_z)$ is uniform over ln(0.1 m) to ln(10 m), and $\ln(\delta^{(t)}_h)$ is uniform over ln(1 m) to ln(100 m). Again, the TMCMC algorithm is adopted to draw posterior samples $\underline{\theta} = \{\underline{\theta}^{(w)},\underline{\theta}^{(t)}\} \sim f(\underline{\theta}|\underline{D})$. The TMCMC algorithm requires the evaluation of the likelihood function, which can be done efficiently through Equation (4.70). The sample size for each TMCMC stage is taken to be N = 2000. In the end of the TMCMC algorithm, 2000 samples of $\underline{\theta} \sim f(\underline{\theta}|\underline{D})$ are obtained. Figure 4.23 shows how the posterior samples of $\underline{\theta}$. The log-evidence $f(\underline{D})$ estimate is 1067.5, much higher than that for the previous example with the quadratic trend (which is 1056.2), suggesting that the GPR method provides a better trend model.

4.2.3.4 Bayesian prediction (conditional random field simulation)

Now consider the simulation of the conditional random field values at the m'_h simulation locations, and, at each location, the random field is simulated at the m_z depths. There are in total $m' = m'_h \times m_z$ locations to simulate. Let $\underline{D}' \in \mathbf{R}^{m'\times 1}$ be formed by stacking the column vectors $\underline{\xi}'^{(1)}, \underline{\xi}'^{(2)}, \ldots,$ and $\underline{\xi}^{(m'h)}$ vertically.

If $\{\underline{\theta}, \underline{t}, \underline{t}'\}$ are known, $\underline{D}' \sim f(\underline{D}'|\underline{\theta},\underline{t},\underline{t}',\underline{D})$ can be readily simulated by Equation (4.56), which is re-written here:

$$\underline{D}' = \underline{t}' + \text{vec}\left[\text{mat}\left(\underline{D} - \underline{t}\right)\mathbf{R}_h^{-1}\mathbf{R}_{h'h}^T\right] + \sigma \times \text{vec}\left[\mathbf{L}_z\text{mat}\left(\mathbf{Z}\right)\mathbf{L}_{h'|h}^T\right] \tag{4.73}$$

where $\mathbf{R}_h \in \mathbf{R}^{mh\times mh}$ is the horizontal auto-correlation matrix of the spatial variability for the m_h sounding locations; $\mathbf{R}_{h'h} \in \mathbf{R}^{mh'\times mh}$ is the horizontal cross-correlation matrix of the spatial variability between the m'_h simulation locations and m_h sounding locations; $\mathbf{L}_z$ is the lower triangular Cholesky decomposition for $\mathbf{R}_z$, and $\mathbf{R}_z$ is the vertical auto-correlation matrix of the spatial variability for the m_z depths; $\mathbf{L}_{h'|h}$ is the lower triangular Cholesky decomposition for $\mathbf{R}_{h'|h}$, and $\mathbf{R}_{h'|h}$ is defined as $\mathbf{R}_{h'} - \mathbf{R}_{h'hRh}{}^{-1}\mathbf{R}_{h'h}{}^{T}$. However, $\{\underline{\theta}, \underline{t}, \underline{t}'\}$ are actually unknown. The purpose of the conditional random field simulation is to draw $\underline{D}'$ samples $\sim f(\underline{D}'|\underline{D})$. According to the Total Probability Theorem and the Law of Large Numbers:

$$f(\underline{D}' \mid \underline{D}) = \int f(\underline{D}' \mid \underline{\theta}, \underline{t}, \underline{t}', \underline{D}) f(\underline{t}, \underline{t}' \mid \underline{\theta}, \underline{D}) f(\underline{\theta} \mid \underline{D}) d\underline{\theta} \approx \frac{1}{N}\sum_{k=1}^{N} f(\underline{D}' \mid \underline{\theta}_k, \underline{t}_k, \underline{t}'_k, \underline{D}) \tag{4.74}$$

where $\underline{\theta}_k$ is the kth TMCMC sample drawn from $f(\underline{\theta}|\underline{D})$, and $\{\underline{t}_k,\underline{t}'_k\}$ are samples drawn from $f(\underline{t},\underline{t}'|\underline{\theta}_k,\underline{D})$. For the GPR method, the sampling of $f(\underline{t},\underline{t}'|\underline{\theta},\underline{D})$ is presented in Appendix 4.2. As a result, the $\underline{D}'$ samples $\sim f(\underline{D}'|\underline{D})$ can be obtained using the following procedure:

1. Randomly draw $\underline{\theta}$ from $\{\underline{\theta}_k$: k = 1, …, N$\}$ and denote this sample by $\hat{\underline{\theta}} = \left\{\hat{\sigma}, \hat{\delta}_h, \hat{\delta}_z, \hat{\nu}_h, \hat{\nu}_z, \hat{\sigma}^{(t)}, \hat{\delta}_h^{(t)}, \hat{\delta}_z^{(t)}\right\}$.
2. Based on $\hat{\underline{\theta}} = \left\{\hat{\sigma}, \hat{\delta}_h, \hat{\delta}_z, \hat{\nu}_h, \hat{\nu}_z, \hat{\sigma}^{(t)}, \hat{\delta}_h^{(t)}, \hat{\delta}_z^{(t)}\right\}$, $\hat{\underline{t}}$ and $\hat{\underline{t}}'$ can be drawn from $f(\underline{t}, \underline{t}' \mid \hat{\underline{\theta}}, \underline{D})$ using the procedure presented in Appendix 4.2. Based on $\left\{\hat{\delta}_h, \hat{\delta}_z, \hat{\nu}_h, \hat{\nu}_z\right\}$, $\{\hat{\mathbf{R}}_h, \hat{\mathbf{R}}_{h'h}, \hat{\mathbf{R}}_z\}$ can be computed, and so are $\{\hat{\mathbf{L}}_{h'|h}, \hat{\mathbf{L}}_z\}$.

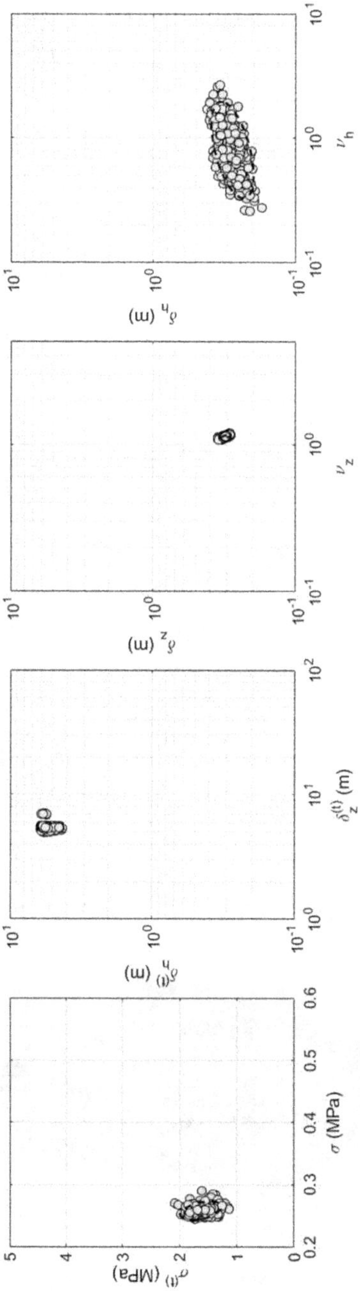

Figure 4.23 Posterior samples of $\underline{\theta}$ (3D spatial data; GPR model).

3. Draw a sample $\underline{D}'$ using the following equation:

$$\underline{D}' = \underline{\hat{t}}' + \text{vec}\left[\text{mat}\left(\underline{D} - \underline{\hat{t}}\right)\hat{\mathbf{R}}_{h}^{-1}\hat{\mathbf{R}}_{h'h}^{T}\right] + \hat{\sigma} \times \text{vec}\left[\hat{\mathbf{L}}_{z}\text{mat}(\mathbf{Z})\hat{\mathbf{L}}_{h'|h}^{T}\right] \tag{4.75}$$

4. Cycle Steps 1–3 to obtain the desirable number of $\underline{D}'$ samples.

Figure 4.24 shows the surface plot of the q_t data and a trend sample $\underline{\hat{t}}$. The trend simulated by the GPR method seems to provide a better fit to the data than the quadratic trend in Figure 4.20. Figure 4.25 shows the conditional random field simulation results at x = 42 m, x = 40 m, x = 40.01 m, and 40.1 m made by the GPR method. The light dots represent one conditional random field sample, the solid line represents the conditional median, and the dashed lines represent the conditional 95% CI. The figure also shows the relevant q_t data for comparison. The dark dots in Figure 4.25a show the actual q_t data at x = 42 m. The dark dots in Figure 4.22b,c,d show the actual q_t data at x = 40 m. Recall that the actual q_t data at x = 42 m are intentionally left out of the training data $\underline{D}$, so the comparison in Figure 4.25a is a genuine validation. With the GPR method, the conditional random field median (solid line) in Figure 4.25a now follows the actual data well. This is in contrast to the poor conditional random field median in Figure 4.22a predicted by the quadratic trend function model. The perfect match between the simulation results and the actual q_t data in Figure 4.25b is correct because the actual q_t data at x = 40 m is within the training data $\underline{D}$. It is remarkable that the 95% CI in Figure 4.25c (x = 40.01 m) is narrow. This is in contrast to the wide 95% CI in Figure 4.22c predicted by the quadratic trend function model. In terms of conditional random field simulation, it is clear that the GPR method (Figure 4.25) performs better than the quadratic trend model (Figure 4.22).

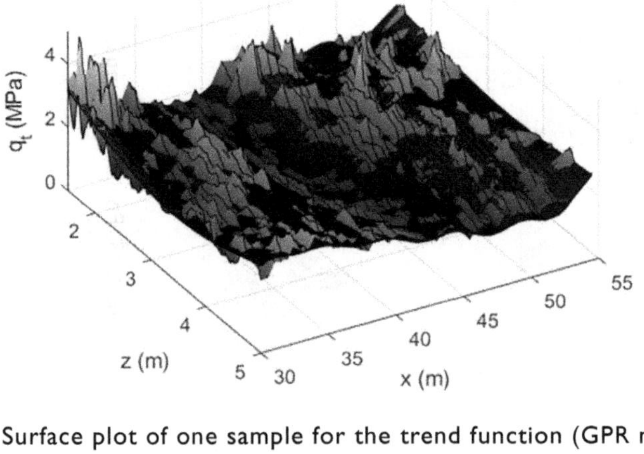

Figure 4.24 Surface plot of one sample for the trend function (GPR method).

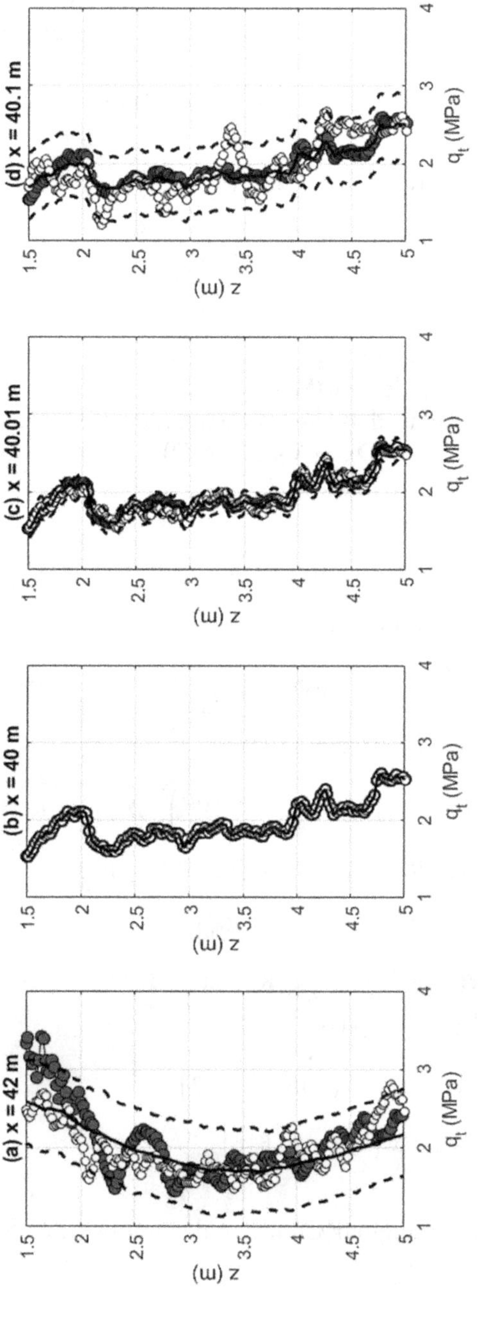

Figure 4.25 Conditional random field simulation results at several locations (GPR method).

APPENDIX 4.1: DERIVATIONS FOR EQUATION (4.70)

This appendix shows that Equation (4.69) can be efficiently evaluated by Equation (4.70) if the following Kronecker-product expressions hold:

$$\mathbf{C} = \sigma^2 \times (\mathbf{R}_h \otimes \mathbf{R}_z) \qquad \mathbf{C}^{(t)} = (\boldsymbol{\Phi}_h \otimes \boldsymbol{\Phi}_z) \times \boldsymbol{\Omega} \times (\boldsymbol{\Phi}_h \otimes \boldsymbol{\Phi}_z)^T \tag{A.1}$$

For the convenience of reference, Equation (4.69) is re-written here:

$$\ln[f(\underline{D} \mid \theta)] = -\frac{m}{2} \times \ln(2\pi) - \frac{1}{2}\ln\left(\left|\mathbf{C}^{(t)} + \mathbf{C}\right|\right) - 0.5 \times \underline{D}^T \left(\mathbf{C}^{(t)} + \mathbf{C}\right)^{-1} \underline{D} \tag{A.2}$$

Let us first focus on the last term $\underline{D}^T(\mathbf{C}^{(t)} + \mathbf{C})^{-1}\underline{D}$ in Equation (4.2). According to the Woodbury matrix identity (Woodbury 1950) $(\mathbf{ABA}^T + \mathbf{C})^{-1} = \mathbf{C}^{-1} - \mathbf{C}^{-1}\mathbf{A}(\mathbf{A}^T\mathbf{C}^{-1}\mathbf{A}+\mathbf{B}^{-1})^{-1}\mathbf{A}^T\mathbf{C}^{-1}$ as well as the following identities: $(\mathbf{A}\otimes\mathbf{B})^{-1} = \mathbf{A}^{-1}\otimes\mathbf{B}^{-1}$, $(\mathbf{A}\otimes\mathbf{B})^T = \mathbf{A}^T\otimes\mathbf{B}^T$, $(\mathbf{AB})\otimes(\mathbf{CD}) = (\mathbf{A}\otimes\mathbf{C})(\mathbf{B}\otimes\mathbf{D})$, $(\mathbf{A}^T\otimes\mathbf{B})\times\text{vec}(\mathbf{C}) = \text{vec}(\mathbf{BCA})$, and $\text{vec}(\mathbf{A})^T\text{vec}(\mathbf{B}) = \text{trace}(\mathbf{A}\times\mathbf{B})$, we have

$$\begin{aligned}
&\underline{D}^T\left(\mathbf{C}^{(t)}+\mathbf{C}\right)^{-1}\underline{D} = \underline{D}^T\left[(\boldsymbol{\Phi}_h\otimes\boldsymbol{\Phi}_z)\boldsymbol{\Omega}(\boldsymbol{\Phi}_h\otimes\boldsymbol{\Phi}_z)^T + \sigma^2(\mathbf{R}_h\otimes\mathbf{R}_z)\right]^{-1}\underline{D} \\
&= \sigma^{-2}\underline{D}^T\Big(\left(\mathbf{R}_h^{-1}\otimes\mathbf{R}_z^{-1}\right) - \left(\mathbf{R}_h^{-1}\otimes\mathbf{R}_z^{-1}\right)(\boldsymbol{\Phi}_h\otimes\boldsymbol{\Phi}_z) \\
&\quad \left[\sigma^2\boldsymbol{\Omega}^{-1} + (\boldsymbol{\Phi}_h\otimes\boldsymbol{\Phi}_z)^T\left(\mathbf{R}_h^{-1}\otimes\mathbf{R}_z^{-1}\right)(\boldsymbol{\Phi}_h\otimes\boldsymbol{\Phi}_z)\right]^{-1}(\boldsymbol{\Phi}_h\otimes\boldsymbol{\Phi}_z)^T\left(\mathbf{R}_h^{-1}\otimes\mathbf{R}_z^{-1}\right)\Big)\underline{D} \\
&= \sigma^{-2}\Big(\underline{D}^T\left(\mathbf{R}_h^{-1}\otimes\mathbf{R}_z^{-1}\right)\underline{D} - \underline{D}^T\left[\left(\mathbf{R}_h^{-1}\boldsymbol{\Phi}_h\right)\otimes\left(\mathbf{R}_z^{-1}\boldsymbol{\Phi}_z\right)\right] \\
&\quad \boldsymbol{\Omega}^{1/2}\left(\sigma^2\mathbf{I} + \boldsymbol{\Omega}^{1/2}\left[\left(\boldsymbol{\Phi}_h^T\mathbf{R}_h^{-1}\boldsymbol{\Phi}_h\right)\otimes\left(\boldsymbol{\Phi}_z^T\mathbf{R}_z^{-1}\boldsymbol{\Phi}_z\right)\right]\boldsymbol{\Omega}^{1/2}\right)^{-1}\boldsymbol{\Omega}^{1/2}\left[\left(\boldsymbol{\Phi}_h^T\mathbf{R}_h^{-1}\right)\otimes\left(\boldsymbol{\Phi}_z^T\mathbf{R}_z^{-1}\right)\right]\underline{D}\Big) \\
&= \sigma^{-2}\Big(\underline{D}^T\text{vec}\left[\mathbf{R}_z^{-1}\text{mat}(\underline{D})\mathbf{R}_h^{-1}\right] - \text{vec}\left[\boldsymbol{\Phi}_z^T\mathbf{R}_z^{-1}\text{mat}(\underline{D})\mathbf{R}_h^{-1}\boldsymbol{\Phi}_h\right]^T \\
&\quad \boldsymbol{\Omega}^{1/2}\left(\mathbf{L}\times\mathbf{L}^T\right)^{-1}\boldsymbol{\Omega}^{1/2}\text{vec}\left[\boldsymbol{\Phi}_z^T\mathbf{R}_z^{-1}\text{mat}(\underline{D})\mathbf{R}_h^{-1}\boldsymbol{\Phi}_h\right]\Big) \\
&= \sigma^{-2}\Big(\underline{D}^T\text{vec}\left[\mathbf{R}_z^{-1}\text{mat}(\underline{D})\mathbf{R}_h^{-1}\right] - \text{vec}\left[\boldsymbol{\Phi}_z^T\mathbf{R}_z^{-1}\text{mat}(\underline{D})\mathbf{R}_h^{-1}\boldsymbol{\Phi}_h\right]^T \\
&\quad \boldsymbol{\Omega}^{1/2}\mathbf{L}^{-T}\mathbf{L}^{-1}\boldsymbol{\Omega}^{1/2}\text{vec}\left[\boldsymbol{\Phi}_z^T\mathbf{R}_z^{-1}\text{mat}(\underline{D})\mathbf{R}_h^{-1}\boldsymbol{\Phi}_h\right]\Big) \\
&= \sigma^{-2}\left(\text{vec}\left[\text{mat}(\underline{D})\right]^T\text{vec}\left[\mathbf{R}_z^{-1}\text{mat}(\underline{D})\mathbf{R}_h^{-1}\right] - \underline{\beta}^T\underline{\beta}\right) \\
&= \sigma^{-2}\left(\text{trace}\left[\text{mat}(\underline{D})^T\mathbf{R}_z^{-1}\text{mat}(\underline{D})\mathbf{R}_h^{-1}\right] - \underline{\beta}^T\underline{\beta}\right)
\end{aligned} \tag{A.3}$$

where $\mathbf{L}$ is the lower triangular Cholesky decompositions for the following matrix:

$$\mathbf{L}\mathbf{L}^T = \sigma^2\mathbf{I} + \boldsymbol{\Omega}^{1/2}\left[\left(\boldsymbol{\Phi}_h^T\mathbf{R}_h^{-1}\boldsymbol{\Phi}_h\right)\otimes\left(\boldsymbol{\Phi}_z^T\mathbf{R}_z^{-1}\boldsymbol{\Phi}_z\right)\right]\boldsymbol{\Omega}^{1/2} \tag{A.4}$$

and $\underline{\beta}$ is defined as

$$\underline{\beta} = \mathbf{L}^{-1}\boldsymbol{\Omega}^{1/2}\text{vec}\left[\boldsymbol{\Phi}_z^T\mathbf{R}_z^{-1}\text{mat}\left(\underline{D}\right)\mathbf{R}_h^{-1}\boldsymbol{\Phi}_h\right] \tag{A.5}$$

Let $(\mathbf{L}_h, \mathbf{L}_z)$ be lower triangular Cholesky decompositions for $(\mathbf{R}_h, \mathbf{R}_z)$, and also based on the identity trace($\mathbf{A}\times\mathbf{B}$) = trace($\mathbf{B}\times\mathbf{A}$), we have

$$\begin{aligned}
&\underline{D}^T\left(\mathbf{C}^{(t)}+\mathbf{C}\right)^{-1}\underline{D} = \sigma^{-2}\left(\text{trace}\left[\text{mat}\left(\underline{D}\right)^T\left(\mathbf{L}_z\mathbf{L}_z^T\right)^{-1}\text{mat}\left(\underline{D}\right)\left(\mathbf{L}_h\mathbf{L}_h^T\right)^{-1}\right]-\underline{\beta}^T\underline{\beta}\right)\\
&= \sigma^{-2}\left(\text{trace}\left[\text{mat}\left(\underline{D}\right)^T\mathbf{L}_z^{-T}\mathbf{L}_z^{-1}\text{mat}\left(\underline{D}\right)\mathbf{L}_h^{-T}\mathbf{L}_h^{-1}\right]-\underline{\beta}^T\underline{\beta}\right)\\
&= \sigma^{-2}\left(\text{trace}\left[\mathbf{L}_h^{-1}\text{mat}\left(\underline{D}\right)^T\mathbf{L}_z^{-T}\times\mathbf{L}_z^{-1}\text{mat}\left(\underline{D}\right)\mathbf{L}_h^{-T}\right]-\underline{\beta}^T\underline{\beta}\right)\\
&= \sigma^{-2}\left(\text{vec}\left[\mathbf{L}_z^{-1}\text{mat}\left(\underline{D}\right)\mathbf{L}_h^{-T}\right]^T\times\text{vec}\left[\mathbf{L}_z^{-1}\text{mat}\left(\underline{D}\right)\mathbf{L}_h^{-T}\right]-\underline{\beta}^T\underline{\beta}\right) = \sigma^{-2}\left(\underline{\alpha}^T\underline{\alpha}-\underline{\beta}^T\underline{\beta}\right)
\end{aligned} \tag{A.6}$$

where

$$\underline{\alpha} = \text{vec}\left[\mathbf{L}_z^{-1}\text{mat}\left(\underline{D}\right)\mathbf{L}_h^{-T}\right] \tag{A.7}$$

Let us now focus on the log-determinant term $\ln(|\mathbf{C}^{(t)}+\mathbf{C}|)$ in Equation (4.2). Applying the matrix determinant identity $|\mathbf{A}+\mathbf{B}\times\mathbf{D}| = |\mathbf{I}+\mathbf{D}\times\mathbf{A}^{-1}\times\mathbf{B}| \times |\mathbf{A}|$, we have

$$\begin{aligned}
&\left|\mathbf{C}^{(t)}+\mathbf{C}\right| = \left|\sigma^2\left(\mathbf{R}_h\otimes\mathbf{R}_z\right)+\left(\boldsymbol{\Phi}_h\otimes\boldsymbol{\Phi}_z\right)\boldsymbol{\Omega}\left(\boldsymbol{\Phi}_h\otimes\boldsymbol{\Phi}_z\right)^T\right|\\
&= \left|\sigma^2\left(\mathbf{R}_h\otimes\mathbf{R}_z\right)+\left(\boldsymbol{\Phi}_h\otimes\boldsymbol{\Phi}_z\right)\boldsymbol{\Omega}^{1/2}\boldsymbol{\Omega}^{1/2}\left(\boldsymbol{\Phi}_h\otimes\boldsymbol{\Phi}_z\right)^T\right|\\
&= \left|\mathbf{I}+\boldsymbol{\Omega}^{1/2}\left(\boldsymbol{\Phi}_h\otimes\boldsymbol{\Phi}_z\right)^T\left[\sigma^2\left(\mathbf{R}_h\otimes\mathbf{R}_z\right)\right]^{-1}\left(\boldsymbol{\Phi}_h\otimes\boldsymbol{\Phi}_z\right)\boldsymbol{\Omega}^{1/2}\right|\times\left|\sigma^2\left(\mathbf{R}_h\otimes\mathbf{R}_z\right)\right|\\
&= \sigma^{2p\times q-2m}\left|\sigma^2\mathbf{I}+\boldsymbol{\Omega}^{1/2}\left[\left(\boldsymbol{\Phi}_h^T\mathbf{R}_h^{-1}\boldsymbol{\Phi}_h\right)\otimes\left(\boldsymbol{\Phi}_z^T\mathbf{R}_z^{-1}\boldsymbol{\Phi}_z\right)\right]\boldsymbol{\Omega}^{1/2}\right|\times\left|\mathbf{R}_h\otimes\mathbf{R}_z\right|
\end{aligned} \tag{A.8}$$

Considering the identity $|\mathbf{A}_{m\times m}\otimes\mathbf{B}_{n\times n}| = |\mathbf{A}|^n|\mathbf{B}|^m$, we have

$$\left|\mathbf{C}^{(t)}+\mathbf{C}\right| = \sigma^{2m-2p\times q}\left|\mathbf{L}\mathbf{L}^T\right|\times\left|\mathbf{R}_h\right|^{m_z}\times\left|\mathbf{R}_z\right|^{m_h} = \sigma^{2m-2p\times q}\left|\mathbf{L}\right|^2\times\left|\mathbf{L}_h\right|^{2m_z}\times\left|\mathbf{L}_z\right|^{2m_h} \tag{A.9}$$

Note that $(\mathbf{L}_h, \mathbf{L}_z, \mathbf{L})$ are all lower triangular matrices, so the determinant is equal to the product of the diagonal terms:

$$\left|\mathbf{C}^{(t)}+\mathbf{C}\right| = \sigma^{2m-2p\times q}\left(\prod_{i=1}^{p\times q}L_{ii}\right)^2\left(\prod_{i=1}^{m_h}L_{h,ii}\right)^{2m_z}\left(\prod_{i=1}^{m_z}L_{z,ii}\right)^{2m_h} \tag{A.10}$$

where $(L_{ii}, L_{h,ii}, L_{z,ii})$ are the ith diagonal terms of the lower triangular matrices $(\mathbf{L}, \mathbf{L}_h, \mathbf{L}_z)$. As a result,

$$\ln\left(\left|C^{(t)}+C\right|\right)=(2m-2p\times q)\ln(\sigma)+2\sum_{i=1}^{p\times q}\ln(L_{ii})+2m_z\sum_{i=1}^{m_h}\ln(L_{h,ii})$$
$$+2m_h\sum_{i=1}^{m_z}\ln(L_{z,ii}) \tag{A.11}$$

Inserting Equations (4.6) and (4.11) into Equation (4.2) yields Equation (4.70).

APPENDIX 4.2: DRAWING $\{\underline{t}, \underline{t}'\}$ SAMPLES FROM $f(\underline{t},\underline{t}'|\underline{\theta},\underline{D})$

Recall from Equation (4.66), $\underline{t}$ can be expressed as

$$\underline{t}=(\boldsymbol{\Phi}_h\otimes\boldsymbol{\Phi}_z)\times\underline{\omega} \tag{A.12}$$

where $\underline{\omega}\in\mathbf{R}^{pq\times1}$ is a zero-mean normal random vector with covariance matrix $=\boldsymbol{\Omega}=\sigma^{(t)2}\times(\boldsymbol{\Omega}_{h,u}\otimes\boldsymbol{\Omega}_z)$. Also, recall that $\underline{D}=\underline{t}+\underline{w}$. It is then clear that $\underline{\omega}$ and $\underline{D}$ are jointly multivariate normal with the following mean vector and covariance matrix:

$$E\left(\begin{bmatrix}\underline{\omega}\\ \underline{D}\end{bmatrix}|\underline{\theta}\right)=\begin{bmatrix}\underline{0}\\ \underline{0}\end{bmatrix} \tag{A.13}$$

$$COV\left(\begin{bmatrix}\underline{\omega}\\ \underline{D}\end{bmatrix}|\underline{\theta}\right)=\begin{bmatrix}\boldsymbol{\Omega} & \boldsymbol{\Omega}(\boldsymbol{\Phi}_h\otimes\boldsymbol{\Phi}_z)^T\\ (\boldsymbol{\Phi}_h\otimes\boldsymbol{\Phi}_z)\boldsymbol{\Omega} & C+(\boldsymbol{\Phi}_h\otimes\boldsymbol{\Phi}_z)\boldsymbol{\Omega}(\boldsymbol{\Phi}_h\otimes\boldsymbol{\Phi}_z)^T\end{bmatrix} \tag{A.14}$$

where $\mathbf{C}=\sigma^2\times(\mathbf{R}_h\otimes\mathbf{R}_z)$ is the covariance matrix of $\underline{w}$. As a result, $f(\underline{\omega}|\underline{\theta},\underline{D})$ is still multivariate normal with the mean vector $\underline{\mu}^{\omega|D}\in\mathbf{R}^{pq\times1}$ and covariance matrix $\mathbf{C}^{\omega|D}\in\mathbf{R}^{pq\times pq}$:

$$\begin{aligned}
\underline{\mu}^{\omega|D}&=E(\underline{\omega}|\underline{\theta},\underline{D})=E(\underline{\omega}|\underline{\theta})+COV(\underline{\omega},\underline{D}|\underline{\theta})COV(\underline{D}|\underline{\theta})^{-1}\left[\underline{D}-E(\underline{D}|\underline{\theta})\right]\\
&=\boldsymbol{\Omega}(\boldsymbol{\Phi}_h\otimes\boldsymbol{\Phi}_z)^T\left[C+(\boldsymbol{\Phi}_h\otimes\boldsymbol{\Phi}_z)\boldsymbol{\Omega}(\boldsymbol{\Phi}_h\otimes\boldsymbol{\Phi}_z)^T\right]^{-1}\\
\underline{D}&=\left[\boldsymbol{\Omega}^{-1}+(\boldsymbol{\Phi}_h\otimes\boldsymbol{\Phi}_z)^TC^{-1}(\boldsymbol{\Phi}_h\otimes\boldsymbol{\Phi}_z)\right]^{-1}(\boldsymbol{\Phi}_h\otimes\boldsymbol{\Phi}_z)^TC^{-1}\underline{D}\\
&=\sigma^{-2}\left[\boldsymbol{\Omega}^{-1}+\sigma^{-2}\left(\boldsymbol{\Phi}_h^TR_h^{-1}\boldsymbol{\Phi}_h\right)\otimes\left(\boldsymbol{\Phi}_z^TR_z^{-1}\boldsymbol{\Phi}_z\right)\right]^{-1}\left[\left(\boldsymbol{\Phi}_h^TR_h^{-1}\right)\otimes\left(\boldsymbol{\Phi}_z^TR_z^{-1}\right)\right]\underline{D}\\
&=\sigma^{-2}\left[\boldsymbol{\Omega}^{-1}+\sigma^{-2}\left(\boldsymbol{\Phi}_h^TR_h^{-1}\boldsymbol{\Phi}_h\right)\otimes\left(\boldsymbol{\Phi}_z^TR_z^{-1}\boldsymbol{\Phi}_z\right)\right]^{-1}\text{vec}\left[\boldsymbol{\Phi}_z^TR_z^{-1}\text{mat}(\underline{D})R_h^{-1}\boldsymbol{\Phi}_h\right]
\end{aligned} \tag{A.15}$$

$$
\begin{aligned}
C^{\omega|D} &= \mathrm{COV}(\underline{\omega}\,|\,\underline{\theta},\underline{D}) = \mathrm{COV}(\underline{\omega}\,|\,\underline{\theta}) - \mathrm{COV}(\underline{\omega},\underline{D}\,|\,\underline{\theta})\mathrm{COV}(\underline{D}\,|\,\underline{\theta})^{-1} \\
&\quad \mathrm{COV}(\underline{\omega},\underline{D}\,|\,\underline{\theta})^{T} \\
&= \Omega - \Omega(\Phi_h \otimes \Phi_z)^T\left[C + (\Phi_h \otimes \Phi_z)\Omega(\Phi_h \otimes \Phi_z)^T\right]^{-1}(\Phi_h \otimes \Phi_z)\Omega \\
&= \left[\Omega^{-1} + (\Phi_h \otimes \Phi_z)^T C^{-1}(\Phi_h \otimes \Phi_z)\right]^{-1} \\
&= \left[\Omega^{-1} + \sigma^{-2}\left(\Phi_h^T R_h^{-1}\Phi_h\right) \otimes \left(\Phi_z^T R_z^{-1}\Phi_z\right)\right]^{-1}
\end{aligned} \tag{A.16}
$$

Note that the evaluations of $\underline{\mu}^{\omega|D}$ and $C^{\omega|D}$ in Equations (4.15) and (4.16) do not require operations (matrix inversion or determinant) of large matrices. Therefore, $\underline{\omega} \sim f(\underline{\omega}|\underline{\theta},\underline{D})$ can be simulated using the following formula with little computation cost:

$$
\begin{aligned}
\underline{\omega} &= \sigma^{-2}\left[\Omega^{-1} + \sigma^{-2}\left(\Phi_h^T R_h^{-1}\Phi_h\right) \otimes \left(\Phi_z^T R_z^{-1}\Phi_z\right)\right]^{-1} \\
&\quad \mathrm{vec}\left[\Phi_z^T R_z^{-1}\mathrm{mat}(\underline{D})R_h^{-1}\Phi_h\right] + L^{\omega|D}\underline{Z}
\end{aligned} \tag{A.17}
$$

where $L^{\omega|D}$ is the lower Cholesky decomposition of $C^{\omega|D}$ in Equation (4.16); $\underline{Z} \in \mathbf{R}^{pq\times 1}$ contains independent standard normal variable samples. Once $\underline{\omega}$ is simulated, $\{\underline{t}, \underline{t}'\}$ samples $\sim f(\underline{t},\underline{t}'|\underline{\theta},\underline{D})$ can be obtained by letting

$$
\begin{aligned}
\underline{t} &= (\Phi_h \otimes \Phi_z) \times \underline{\omega} = \mathrm{vec}\left[\Phi_z \times \mathrm{mat}(\underline{\omega}) \times \Phi_h^T\right] \\
\underline{t}' &= (\Phi_{h'} \otimes \Phi_z) \times \underline{\omega} = \mathrm{vec}\left[\Phi_z \times \mathrm{mat}(\underline{\omega}) \times \Phi_{h'}^T\right]
\end{aligned} \tag{A.18}
$$

Chapter 5

Full-scale real case study

The purpose of this chapter is to address some considerations in practice. Realistic geotechnical site investigation data are multivariate, unique, sparse, incomplete, potentially corrupted, and spatially variable. Phoon (2020) and Ching et al. (2022) used the term "MUSIC-3X" to describe the common attributes of real-world site investigation data. The site investigation data are multivariate because multiple tests (e.g., Atterberg's limit test, cone penetration test, and triaxial compression test) may be used to characterize a site, unique because of the site uniqueness, sparse because measurements are taken at limited locations, incomplete because not all tests are conducted at each location and depth, potentially corrupted because outliers may exist, and spatially variable because soil properties intrinsically vary in space. The analysis methods in previous chapters cannot fully address all these realistic MUSIC-3X aspects. In this chapter, a full-scale real case study (Baytown site) is analyzed. During the analyses, useful algorithms are developed to address these realistic MUSIC-3X aspects.

5.1 REAL CASE STUDY – BAYTOWN SITE

To illustrate a realistic full-scale example, a test site at Baytown, Texas, USA (Stuedlein et al. 2012) is adopted in this section. The site exploration plan is shown in Figure 5.1. The soil at the test site is mainly clay (known as Beaumont clay) with occasional layers of silt and silty fine sand. There is a desiccated very stiff clay crust underlain by an upper Beaumont clay layer (medium stiff), a thin silty sand and sandy silt layer, and a lower Beaumont clay layer (stiff to very stiff). The ground water table is at the depth of 2.9 m.

The site investigation program includes the following:

1. Nine cone penetration test (CPT) soundings (CPT-1 to -3 and CPT-F1 to -F6, as shown in Figure 5.2) to determine cone resistance (q_t), sleeve friction (f_s), and penetration-induced pore water pressure (u_2). The normalized cone resistance and pore water pressure parameter can be further calculated as $q_{t1} = (q_t - \sigma_v)/\sigma'_v$ and $B_q = (u_2 - u_0)/(q_t - \sigma_v)$,

DOI: 10.1201/9781003309765-5

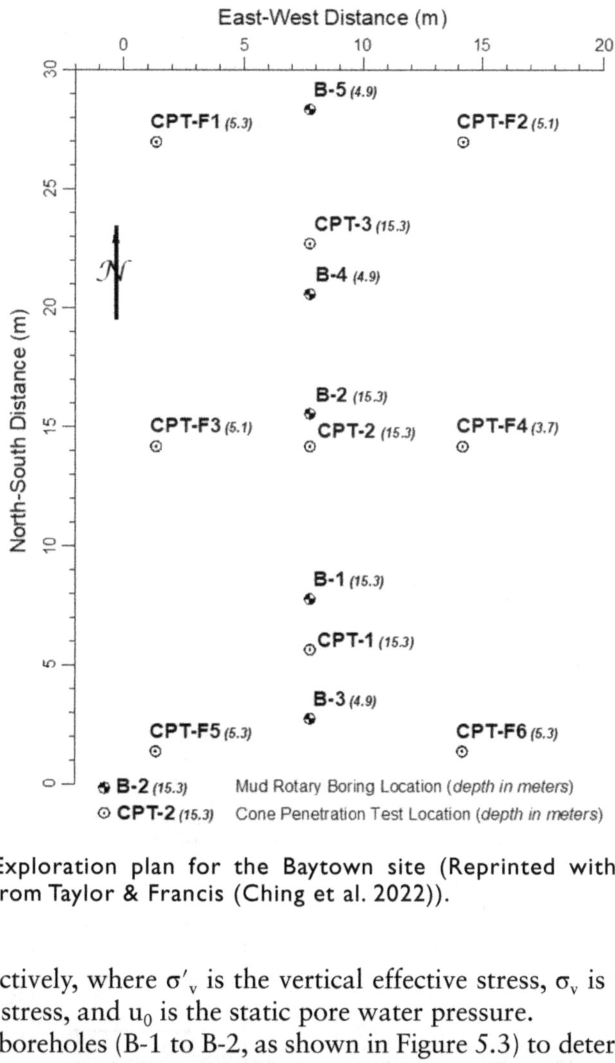

Figure 5.1 Exploration plan for the Baytown site (Reprinted with permission from Taylor & Francis (Ching et al. 2022)).

respectively, where σ'_v is the vertical effective stress, σ_v is the vertical total stress, and u_0 is the static pore water pressure.

2. Two boreholes (B-1 to B-2, as shown in Figure 5.3) to determine liquid limit (LL), plastic limit (PL), and natural water content (w). They can be further used to calculate plasticity index (PI) and liquidity index (LI).
3. Three boreholes (B-3 to B-5, as shown in Figure 5.4) to determine the preconsolidation stress (σ'_p) and undrained shear strength (s_u). The index properties (LL, PL, w) for the samples from B-3 to B-5 are also known.

To further illustrate the site investigation data in Figures 5.2–5.4, Tables 5.1–5.3 show the data for B-1, B-3, and CPT-1 as examples. Each table represents one borehole/sounding. There are n = 8 soil parameters of interest, denoted by $(Y_1, Y_2, \ldots, Y_8) = (\ln(LL), \ln(PI), LI, \ln(\sigma'_v/P_a), \ln(OCR), \ln(s_u/\sigma'_v), B_q, \ln(q_{t1}))$ ($P_a = 101.3$ kN/m^2 is one atmosphere pressure). There are in total

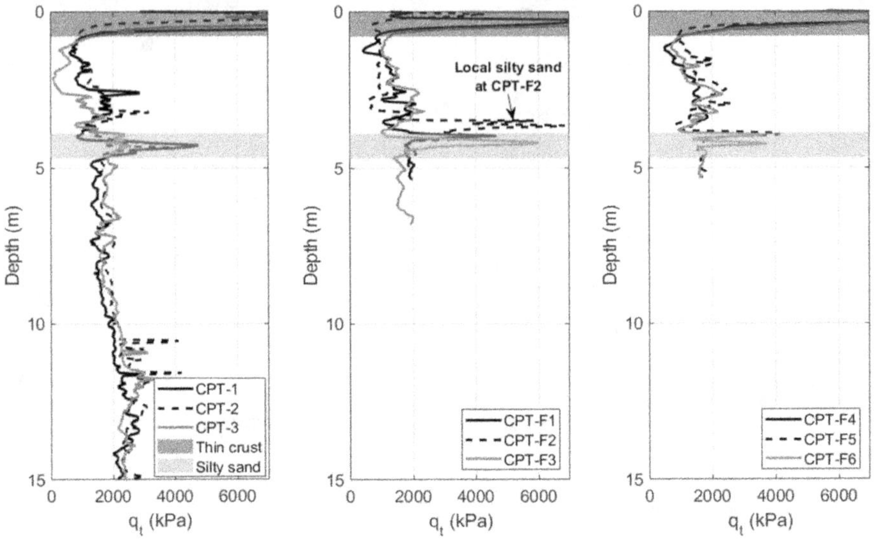

Figure 5.2 CPT test results (Reprinted with permission from Taylor & Francis (Ching et al. 2022)).

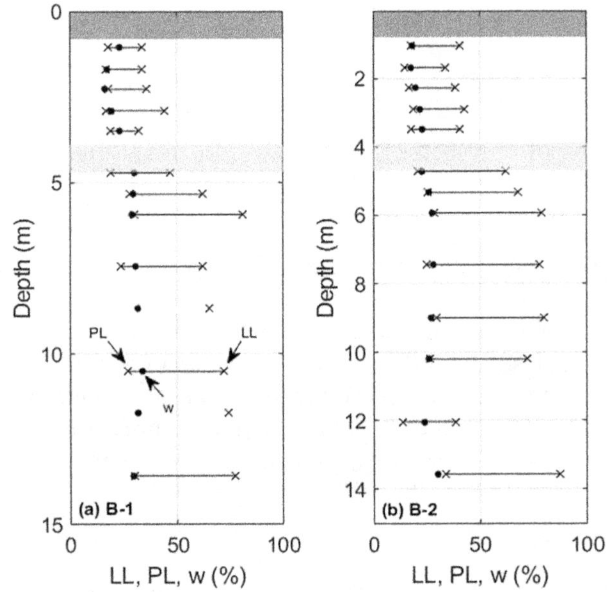

Figure 5.3 Test results for B-1 and B-2 (Reprinted with permission from Taylor & Francis (Ching et al. 2022)).

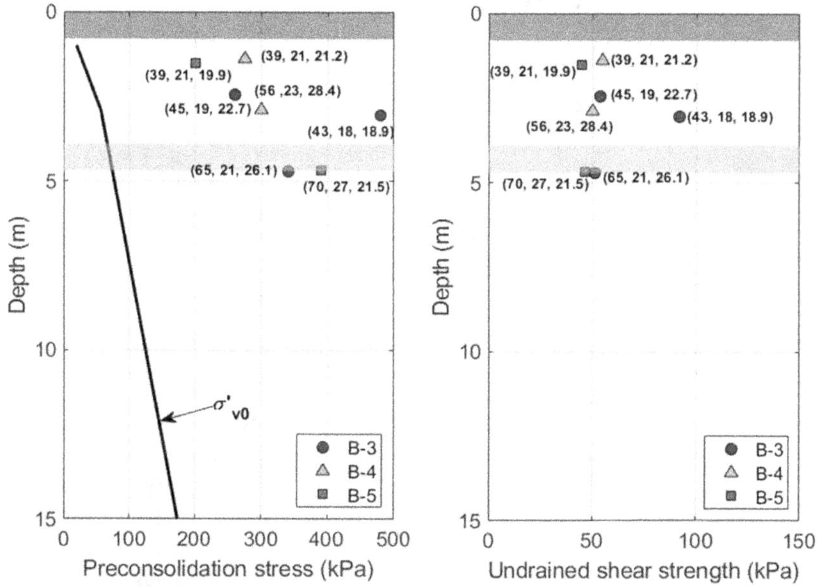

Figure 5.4 Test results for B-3 to B-5: (a) preconsolidation stress; (b) undrained shear strength. The index properties (LL, PL, w) are annotated (Reprinted with permission from Taylor & Francis (Ching et al. 2022)).

Table 5.1 Site investigation data at B-1

Depth (m)	*Index properties*			*Stresses*		*Strength*	*CPT*	
	LL (%)	*PI (%)*	*LI*	σ'_v/P_a	*OCR*	s_u/σ'_v	B_q	q_{t1}
1.06	34	16	0.34	0.20				
1.68	34	17	0.05	0.32				
2.28	36	18	−0.08	0.44				
2.90	44	27	0.10	0.56				
4.72	47	28	0.39	0.73				
5.34	62	34	0.04	0.79				
5.94	81	51	−0.01	0.85				
7.46	62	38	0.17	0.99				
8.68	65			1.11				
10.52	72	45	0.15	1.29				
11.72	74			1.40				
13.56	77	47		1.58				
15.08	81	51		1.72				

Source: Ching et al. 2022.

Table 5.2 Site investigation data at B-3

Depth (m)	*Index properties*			*Stresses*		*Strength*	*CPT*	
	LL (%)	*PI (%)*	*LI*	σ'_v/P_a	*OCR*	s_u/σ'_v	B_q	q_{tl}
2.44	45	26	0.14	0.47	5.46	1.13[a]		
3.04	43	25	0.04	0.57	8.29	1.59[a]		
4.72	65	44	0.12	0.73	4.58	0.69[a]		

Source: Ching et al. 2022.

[a] s_u evaluated by the SHANSEP equation based on σ'_p.

Table 5.3 Site investigation data at CPT-1

Depth (m)	*Index properties*			*Stresses*		*Strength*	*CPT*	
	LL (%)	*PI (%)*	*LI*	σ'_v/P_a	*OCR*	s_u/σ'_v	B_q	q_{tl}
1.0				0.19			−0.0012	42.63683
1.1				0.21			−0.0026	35.03541
1.2				0.23			−0.0095	30.02517
1.3				0.25			−0.0063	23.9831
…								
3.3				0.60			−0.035	27.86
3.4				0.61			−0.047	20.67
4.7				0.73			−0.032	22.31
4.8				0.74			−0.039	19.34
…								
9.9				1.23			−0.021	14.34
10.0				1.24			−0.022	13.34

Source: Ching et al. 2022.

m_h = 14 boreholes/soundings (five boreholes and nine CPTs; not shown in Tables 5.1–5.3 for brevity). The number and distance between borehole samples are fairly sparse. There are 13 sample depths for B-1 and only 3 sample depths for B-3, and different boreholes implemented samples at different depths. The union of the observation sample depths for all five boreholes is called the "borehole sample depths". There are 21 borehole sample depths.

The number of data samples obtained from the CPTs is dense with a sample interval of 0.02 m. To reduce the computational cost, all CPT data are re-sampled at a depth interval of 0.1 m. The 9 CPTs also extend to different depths (see Figure 5.2). The deepest CPT explores a maximum depth of 15.32 m. In this section, the CPT data are analyzed to a maximum depth of 10 m. The CPT data shallower than 1 m are not analyzed because of the thin crust near the ground surface. Also, recall that there is a thin silty sand layer (see Figure 5.2). As a result, the CPT data between 3.4 and 4.7 m are not analyzed either, because of the thin silty sand layer. The union of the sample depths for all nine CPTs is called the "CPT sample depths". There are 79 CPT sample depths. The borehole/sounding sample depths for the Baytown site is the union of the 21 borehole sample depths and the 79 CPT sample depths. There are $m_z = 98$ borehole/sounding sample depths (there are 2 overlapping sample depths). The data in Tables 5.1–5.3 clearly do not follow a lattice structure. To facilitate a lattice representation of this data, these three tables are expanded into larger tables. For instance, Table 5.1 is expanded into a table with $m_z = 98$ depths, and most of the entries are empty (except σ'_v/P_a which is generally applicable to all available data). After the expansion, the site investigation data follow a lattice structure. It is clear that the (expanded) data are sparse (limited boreholes and soundings) and incomplete (characterized with missing entries). In total, there are $m = m_h \times m_z = 1372$ lattice locations.

5.2 MODELING OF THE BAYTOWN-SITE DATA

Let us denote the (transformed) soil parameter at the jth lattice location at the Baytown site by $\underline{X}_s^{(j)} \in \mathbf{R}^{n\times 1}$ (n = 8). The transformation that converts $(Y_1, Y_2, \ldots, Y_8)$ to $(X_1, X_2, \ldots, X_8)$ will be discussed later. A key assumption made is that $\underline{X}_s^{(j)}$ follows the multivariate normal PDF $N(\underline{x}; \underline{\mu}_s, \mathbf{C}_s)$, where $\underline{\mu}_s \in \mathbf{R}^{n\times 1}$ is the site-specific mean vector of the Baytown site, and $\mathbf{C}_s \in \mathbf{R}^{n\times n}$ is the site-specific covariance matrix. Let $\mathbf{X}_s = \{\underline{X}_s^{(1)}, \underline{X}_s^{(2)}, \ldots, \underline{X}_s^{(m)}\}$ be the collection of all Baytown-site data, and let $\underline{X}_s$ be an $[(n\times m)\times 1]$ column vector formed by stacking the column vectors $\underline{X}_s^{(1)}, \underline{X}_s^{(2)}, \ldots$, and $\underline{X}_s^{(m)}$ vertically. As a result, $\underline{X}_s$ is distributed as the following multivariate normal PDF:

$$\underline{X}_s \sim f(\underline{X}_s \mid \underline{\mu}_s, \mathbf{C}_s) = N\left(\underline{X}_s; \underline{1}_{m\times 1} \otimes \underline{\mu}_s, \mathbf{R} \otimes \mathbf{C}_s\right)$$
$$= N\left(\underline{X}_s; \begin{bmatrix} \underline{\mu}_s \\ \underline{\mu}_s \\ \vdots \\ \underline{\mu}_s \end{bmatrix}, \begin{bmatrix} \mathbf{C}_s & \rho_{12}\mathbf{C}_s & \cdots & \rho_{1m}\mathbf{C}_s \\ & \mathbf{C}_s & & \rho_{2m}\mathbf{C}_s \\ & & \ddots & \vdots \\ \text{SYM.} & & & \mathbf{C}_s \end{bmatrix}\right) \tag{5.1}$$

where $\underline{1}_{m\times 1} \in \mathbf{R}^{m\times 1}$ denotes an (m×1) vector containing ones; $\mathbf{R} \in \mathbf{R}^{m\times m}$ is the auto-correlation matrix that prescribes the spatial-correlation structure:

$$\mathbf{R} = \begin{bmatrix} 1 & \rho_{12} & \rho_{13} & \cdots & \rho_{1m} \\ & 1 & \rho_{23} & \cdots & \rho_{2m} \\ & & 1 & & \\ & & & \ddots & \vdots \\ \text{SYM.} & & & & 1 \end{bmatrix} \tag{5.2}$$

where ρ_{ij} is the correlation coefficient between the ith and jth locations. Note that by expressing the covariance matrix as $\mathbf{R}\otimes\mathbf{C}_s$ in Equation (5.1), the separability assumption between the spatial correlation (**R** matrix) and cross-correlation ($\mathbf{C}_s$ matrix) is adopted, that is, all soil parameters follow the same auto-covariance parameters. Moreover, the separability assumption between the horizontal and vertical directions is adopted:

$$\mathbf{R} = \mathbf{R}_h \otimes \mathbf{R}_z \tag{5.3}$$

where $\mathbf{R}_h \in \mathbf{R}^{mh\times mh}$ is the horizontal auto-correlation matrix among the m_h sounding locations, and $\mathbf{R}_z \in \mathbf{R}^{mz\times mz}$ is the vertical auto-correlation matrix among the n_z sampling depths.

5.3 CONSTRUCTION OF INFORMATIVE PRIOR USING HBM – DEALING WITH MISSING DATA IN THE DATABASE

The HBM is adopted to construct the informative prior model of the Baytown site-specific parameters ($\underline{\mu}_s$, $\mathbf{C}_s$). The CLAY/10/7490 database collected by Ching and Phoon (2014) contains (Y_1, ..., Y_8) data from more than 200 sites, and this database (denoted by $\mathbf{X}^o$; the superscript "o" denotes "observed") is adopted to train the HBM for the site-uniqueness characteristics in $\mathbf{X}^o$. First, the Johnson system of distributions is adopted to model the marginal PDFs of the (Y_1, ..., Y_8) data in $\mathbf{X}^o$. Once the Johnson parameters (including the distribution type and η, γ, λ, ε) are fitted, the (Y_1, ..., Y_8) data are transformed into (X_1, ..., X_8).

However, the database $\mathbf{X}^o$ is an incomplete multivariate database. For many cases in $\mathbf{X}^o$, (X_1, ..., X_8) are not completely observed, and there are missing entries. The collection of all missing entries is denoted by $\mathbf{X}^u$ (the superscript "u" denotes "unobserved"). This missing data issue can be resolved because the missing data can be readily simulated during the Gibbs sampler (GS) procedure in the HBM. To elaborate, let us denote the data at the jth location of the ith site in $\mathbf{X}^o$ by $\underline{X}_i^{(j)}$. Let us further denote missing (un-observed) values in $\underline{X}_i^{(j)}$ by $\underline{X}_i^{(j)u}$ and denote the observed values by $\underline{X}_i^{(j)o}$. Let the mean vector $\underline{\mu}_i$ and covariance matrix $\mathbf{C}_i$ for the ith site in the database are both partitioned accordingly: $\underline{\mu}_i^u$ is the sub-mean vector in $\underline{\mu}_i$ for the un-observed components, and $\underline{\mu}_i^o$ is the sub-mean vector in $\underline{\mu}_i$ for the observed components. Similarly, $\mathbf{C}_i^u$ is the sub-square matrix in $\mathbf{C}_i$ for the un-observed components, $\mathbf{C}_i^o$ is the sub-square matrix in $\mathbf{C}_i$ for the observed components, and $\mathbf{C}_i^{uo}$ is the sub-covariance matrix in $\mathbf{C}_i$ for the covariance between $\underline{X}_i^{(j)u}$ and $\underline{X}_i^{(j)o}$. It can be shown that the fully conditional PDF of $\underline{X}_i^{(j)u}$ during the GS is still multivariate normal:

$$\begin{aligned} &f(\underline{X}_i^{(j)u} \mid \underline{X}_i^{(j)o}, \mu, C, \underline{\mu}_0, C_0, \nu_0, \Sigma_0, \underline{a}) \\ &= N\left\{ \underline{\mu}_i^u + C_i^{uo}\left(C_i^o\right)^{-1}\left(\underline{X}_i^{(j)o} - \underline{\mu}_i^o\right), C_i^u - C_i^{uo}\left(C_i^o\right)^{-1} C_i^{ou} \right\} \end{aligned} \tag{5.4}$$

As a result, it is possible to simulate the missing values $\underline{X}_i^{(j)u}$ during the GS. This step of simulating missing values can be readily augmented into the HBM GS procedure as Step 9 in the following:

1. Initialize $(\mathbf{X}^u,\boldsymbol{\mu},\mathbf{C},\boldsymbol{\Theta},\underline{a})$ samples at arbitrary values.
2. (For i = 1, ..., n_s): Draw $\underline{\mu}_i$ sample ~ $f(\underline{\mu}_i|\boldsymbol{\mu}_{\backslash i},\mathbf{C},\boldsymbol{\Theta},\underline{a})$:

$$\underline{\mu}_i \sim N\left\{\left(C_0^{-1}+m_iC_i^{-1}\right)^{-1}\left(C_0^{-1}\underline{\mu}_0+C_i^{-1}\sum_{j=1}^{m_i}\underline{X}_i^{(j)}\right),\left(C_0^{-1}+m_iC_i^{-1}\right)^{-1}\right\} \tag{5.5}$$

3. (For i = 1, ..., n_s): Draw C_i sample ~ $f(C_i|\boldsymbol{\mu},C_{\backslash i},\boldsymbol{\Theta},\underline{a})$:

$$C_i \sim IW\left\{\boldsymbol{\Sigma}_0+\sum_{j=1}^{m_i}\left(\underline{X}_i^{(j)}-\underline{\mu}_i\right)\left(\underline{X}_i^{(j)}-\underline{\mu}_i\right)^T, m_i+\nu_0\right\} \tag{5.6}$$

4. Draw $\underline{\mu}_0$ sample ~ $f(\underline{\mu}_0|\boldsymbol{\mu},\mathbf{C},C_0,\boldsymbol{\Sigma}_0,\nu_0,\underline{a})$:

$$\underline{\mu}_0 \sim N\left\{\left(C_{\mu 0}^{-1}+n_sC_0^{-1}\right)^{-1}\left(C_{\mu 0}^{-1}\underline{\mu}_{\mu 0}+C_0^{-1}\sum_{i=1}^{n_s}\underline{\mu}_i\right),\left(C_{\mu 0}^{-1}+n_sC_0^{-1}\right)^{-1}\right\} \tag{5.7}$$

5. Draw C_0 sample ~ $f(C_0|\boldsymbol{\mu},\mathbf{C},\underline{\mu}_0,\boldsymbol{\Sigma}_0,\nu_0,\underline{a})$:

$$C_0 \sim IW\left\{\boldsymbol{\Sigma}_{C0}+\sum_{i=1}^{n_s}\left(\underline{\mu}_i-\underline{\mu}_0\right)\left(\underline{\mu}_i-\underline{\mu}_0\right)^T, n_s+\nu_{C0}\right\} \tag{5.8}$$

6. Draw $\boldsymbol{\Sigma}_0$ sample ~ $f(\boldsymbol{\Sigma}_0|\boldsymbol{\mu},\mathbf{C},\underline{\mu}_0,C_0,\nu_0,\underline{a})$:

$$\boldsymbol{\Sigma}_0 \sim W\left\{\left(\boldsymbol{\Psi}_{\Sigma 0}^{-1}+\sum_{i=1}^{n_s}C_i^{-1}\right)^{-1}, n_s\nu_0+\lambda_{\Sigma 0}\right\} \tag{5.9}$$

7. Draw ν_0 sample from the following discrete distribution:

$$p(\nu_0 \mid \mathbf{X},\boldsymbol{\mu},\mathbf{C},\underline{\mu}_0,C_0,\boldsymbol{\Sigma}_0,\underline{a}) \propto \frac{\prod_{i=1}^{n_s}\left|C_i\right|^{-\frac{\nu_0}{2}}\cdot\left|\boldsymbol{\Sigma}_0\right|^{\frac{n_s\nu_0}{2}}}{2^{n_s n\nu_0/2}\cdot\Gamma_n\left(\nu_0/2\right)^{n_s}} \quad \nu_0=n,n+1,\ldots,1000 \tag{5.10}$$

8. (For k = 1, ..., n): Draw a_k sample ~ $f(a_k|\boldsymbol{\mu},\mathbf{C},\underline{\mu}_0,\mathbf{C}_0,\boldsymbol{\Sigma}_0,\nu_0,\underline{a}_{\backslash k})$:

$$a_k \sim IG\left(\alpha + \frac{\nu_{C_0}}{2}, \beta + 2C_{0,kk}^{-1}\right) \tag{5.11}$$

9. (For i = 1, ..., n_s; j = 1, ..., m_i): Draw $\underline{X}_i^{(j)u}$ sample ~ $f(\underline{x}_i^{(j)u}|\underline{x}_i^{(j)o},\boldsymbol{\mu},\mathbf{C},\underline{\mu}_0,\mathbf{C}_0,\boldsymbol{\Sigma}_0,\underline{a})$:

$$\underline{X}_i^{(j)u} \sim N\left\{\underline{\mu}_i^u + C_i^{uo}\left(C_i^o\right)^{-1}\left(\underline{X}_i^{(j)o} - \underline{\mu}_i^o\right), C_i^u - C_i^{uo}\left(C_i^o\right)^{-1}C_i^{ou}\right\} \tag{5.12}$$

10. Cycle Steps 2–9 for T time steps to obtain T samples for $(\mathbf{X}^u,\boldsymbol{\mu},\mathbf{C},\boldsymbol{\Theta},\underline{a})$.

The HBM main outcome is the hyper-parameter samples $\{\boldsymbol{\Theta}_k$: k = 1, ..., N}, where N = $(T\text{-}T_b)/\Delta t$, that are distributed as $f(\boldsymbol{\Theta}|\mathbf{X}^o)$. They have been trained by the soil database $\mathbf{X}^o$ about the site-uniqueness characteristics.

To illustrate the HBM trained by $\mathbf{X}^o$, Figure 5.5 shows the distribution of site statistics for the X_5–X_6 relation [X_5 is transformed from ln(OCR), and X_6 is transformed from $\ln(s_u/\sigma'_v)$]. For each site in CLAY/10/7490 with (X_5, X_6) data, the sample means of the (X_5, X_6) are denoted by (m_5, m_6), sample standard deviations denoted by (s_5, s_6), and sample correlation denoted by r_{56}. There are many such sites with (X_5, X_6) data in CLAY/10/7490, the distributions of the statistics are shown in Figure 5.5 as the circles and dark histogram. They represent the actual site statistics in CLAY/10/7490. On the other hand, the HBM training outcome is $\{\boldsymbol{\Theta}_k$: k = 1, ..., N}, where $\boldsymbol{\Theta}$ contains the hyper-parameters $\{\underline{\mu}_0,\mathbf{C}_0,\boldsymbol{\Sigma}_0,\nu_0\}$. Recall that $(\underline{\mu}_0,\mathbf{C}_0)$ governs the prior PDF of $\underline{\mu}_s$, and $(\boldsymbol{\Sigma}_0,\nu_0)$ governs the prior PDF of $\mathbf{C}_s$. As mentioned in Section 4.1.3, the trained HBM can produce the prior PDF $f(\underline{\mu}_s,\mathbf{C}_s|\mathbf{X}^o)$ for the Baytown site via the hyper-parameter samples. Figure 5.5 illustrates this prior PDF in its fifth component [X_5 = transformed ln(OCR)] and sixth component [X_6 = transformed $\ln(s_u/\sigma'_v)$]. The contour lines and light histogram show the probability distributions of $\{\mu_5, \mu_6\}$, $\{\sigma_5, \sigma_6\}$ and ρ_{56} of the trained HBM. It is evident from Figure 5.5 that the distributions for $\{m_5, m_6, s_5, s_6, r_{56}\}$ and $\{\mu_5, \mu_6, \sigma_5, \sigma_6, \rho_{56}\}$ are similar, suggesting that the HBM has consistently learned the OCR-s_u/σ'_v cross-correlation in CLAY/10/7490. The illustration in Figure 5.5 is for the OCR-s_u/σ'_v relation. Although not shown, similar consistency was observed for other pairwise relations.

5.4 ESTIMATION OF AUTO-CORRELATIONS – DEALING WITH NON-LATTICE CPT DATA

To address the spatial correlation of the target site, it is necessary to estimate the auto-correlation matrices $\mathbf{R}_h$ and $\mathbf{R}_z$ based on the Baytown site data. For this purpose, the CPT $\ln(q_{t1})$ data (denoted by $\boldsymbol{\xi}$ in this section) in the upper

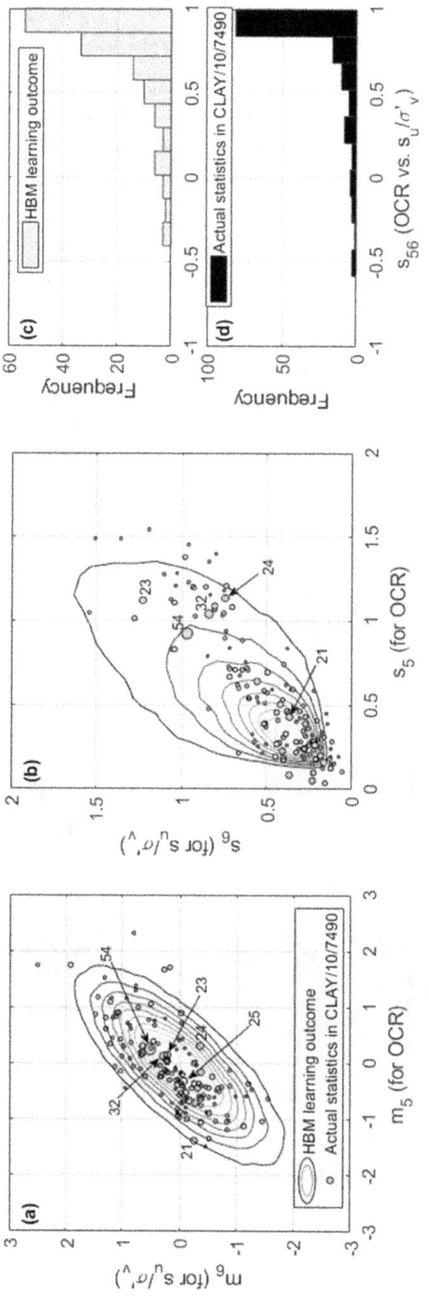

Figure 5.5 Site statistics for (a) m_5 & m_6, (b) s_5 & s_6, and (c) r_{56}. The size of the circle indicates the amount of x_5–x_6 data (cases with more than 20 data points > 20 are annotated with #) (Reprinted with permission from Taylor & Francis (Ching et al. 2022)).

and lower Beaumont clays is adopted to estimate the site-specific auto-correlation parameters. The GPR method in Section 4.2.3 is adopted: the trend part is modeled by the QExp auto-correlation model, whereas the spatial variability part is modeled by the WM model. There are four parameters for the WM model: vertical SOF (δ_z), horizontal SOF (δ_h), vertical smoothness (ν_z), and horizontal smoothness (ν_h). As mentioned in Section 4.2.3, if the CPT data follows a lattice structure, the computation for the likelihood function in Equations (4.70)–(4.72) is fast even for 3D problems due to the Kronecker-product derivations in Section 4.2.3. However, the CPTs at the Baytown site do not strictly follow a lattice structure because they do not measure identical depths (see Figure 5.2). Nevertheless, if the non-lattice data contain a lattice subset (as elaborated soon), it is still possible to take advantage of the Kronecker-product derivations.

Figure 5.6 illustrates a set of non-lattice data from m_h soundings, where not all soundings measure identical depths. The data in Figure 5.6 are divided into two subsets. The lattice subset $\xi^{lat} \in \mathbf{R}^{(m_z^{lat} \times m_h)\times 1}$ is within the two dashed boxes, occupying m_z^{lat} depths (denoted by the lattice depths). The subset $\xi^{\sim lat} \in \mathbf{R}^{[(m_z - m_z^{lat})\times m_h]\times 1}$ is outside the boxes, occupying the remaining $(m_z - m_z^{lat}) = m_z^{\sim lat}$ depths (denoted by the non-lattice depths). Let $\xi^{\sim lat}$ be further divided into two subsets: the observed subset $\xi^o \in \mathbf{R}^{m^o \times 1}$ and unobserved (missing) subset $\xi^u \in \mathbf{R}^{m^u \times 1}$, where m^o and m^u are the numbers of

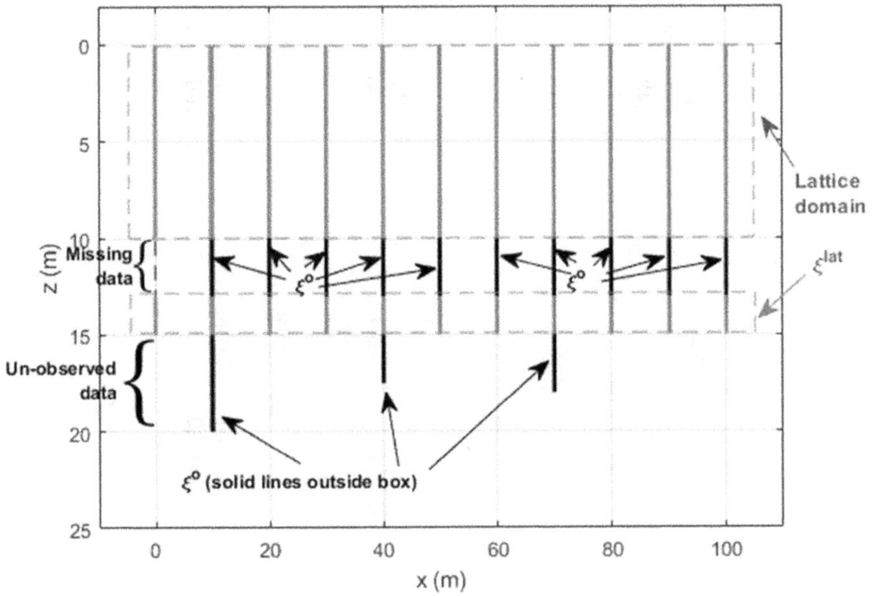

Figure 5.6 Illustration for non-lattice CPT data (Reprinted with permission from American Society of Civil Engineers (Ching et al. 2021b)).

observed and un-observed data points in $\underline{\xi}^{\sim lat}$, respectively (note: $m^o + m^u = m_z^{\sim lat} \times m_h$). As a result, the observed CPT data contains $\underline{\xi}^o$ and $\underline{\xi}^{lat}$.

To implement the GPR method, it is required to evaluate the log-likelihood function $\ln[f(\underline{\xi}^o,\underline{\xi}^{lat}|\underline{\theta})]$. Because $\{\underline{\xi}^o,\underline{\xi}^{lat}\}$ is non-lattice, Equations (4.70)–(4.72) cannot be adopted to achieve fast computation. Nonetheless, the derivations in this section will show that it is still possible to evaluate $\ln[f(\underline{\xi}^o,\underline{\xi}^{lat}|\underline{\theta})]$ efficiently by exploring the lattice structure in $\underline{\xi}^{lat}$. The notations used in this section follow the GPR section (Section 4.2.3):

1. $\mathbf{R}_h$ and $\mathbf{R}_z$ denote the horizontal and vertical auto-correlation matrices for the spatial variability. Furthermore, let us denote $\mathbf{R}_z^{lat} \in \mathbf{R}^{m_z^{lat} \times m_z^{lat}}$ and $\mathbf{R}_z^{\sim lat} \in \mathbf{R}^{m_z^{\sim lat} \times m_z^{\sim lat}}$ as the sub-matrices of $\mathbf{R}_z$ for the auto-correlation of the lattice and non-lattice depths, respectively, and $\mathbf{R}_z^{\sim lat,lat} \in \mathbf{R}^{m_z^{\sim lat} \times m_z^{lat}}$ as the sub-matrix of $\mathbf{R}_z$ for the cross-correlation between the non-lattice and lattice depths.
2. $\mathbf{\Phi}_h \in \mathbf{R}^{mh \times p}$ and $\mathbf{\Phi}_z \in \mathbf{R}^{mz \times q}$ contain the eigenvectors of the horizontal and vertical, respectively, auto-correlation matrices for the trend (p and q are the numbers of eigenvectors adopted in Equations 4.61 and 4.63). Let us further denote $\mathbf{\Phi}_z^{lat} \in \mathbf{R}^{m_z^{lat} \times mz}$ and $\mathbf{\Phi}_z^{\sim lat} \in \mathbf{R}^{m_z^{\sim lat} \times mz}$ as the sub-matrices of $\mathbf{\Phi}_z$ for the lattice and non-lattice depths, respectively.

To start the derivations, it is first noted that

$$\ln\left[f(\underline{\xi}^o,\underline{\xi}^{lat} \mid \underline{\theta})\right] = \ln\left[f(\underline{\xi}^o \mid \underline{\theta},\underline{\xi}^{lat})\right] + \ln\left[f(\underline{\xi}^{lat} \mid \underline{\theta})\right] \tag{5.13}$$

The term $\ln[f(\underline{\xi}^{lat}|\underline{\theta})]$ can be efficiently evaluated by the following equations similar to Equations (4.70)–(4.72) because $\underline{\xi}^{lat}$ contains lattice data:

$$\ln\left[f(\underline{\xi}^{lat} \mid \underline{\theta})\right] = -\frac{m_z^{lat} m_h}{2}\ln(2\pi) - m_z^{lat}\sum_{i=1}^{m_h}\ln\left(L_{h,ii}\right) - m_h\sum_{i=1}^{m_z^{lat}}\ln\left(L_{z,ii}^{lat}\right)$$
$$-\left(m_z^{lat}m_h - p\times q\right)\times\ln(\sigma) - \sum_{i=1}^{p\times q}\ln\left(L_{ii}^{lat}\right) - \frac{1}{2\sigma^2}\left[\left(\underline{\alpha}^{lat}\right)^T\underline{\alpha}^{lat} - \left(\underline{\beta}^{lat}\right)^T\underline{\beta}^{lat}\right] \tag{5.14}$$

where $(\mathbf{L}_h,\mathbf{L}_z^{lat})$ are lower triangular Cholesky decompositions for $(\mathbf{R}_h,\mathbf{R}_z^{lat})$;

$$\underline{\alpha}^{lat} = \text{vec}\left[\left(\mathbf{L}_z^{lat}\right)^{-1}\text{mat}\left(\underline{\xi}^{lat}\right)\mathbf{L}_h^{-T}\right] \quad \underline{\beta}^{lat} = \left(\mathbf{L}^{lat}\right)^{-1}$$
$$\times\mathbf{\Omega}^{1/2}\times\text{vec}\left[\left(\mathbf{\Phi}_z^{lat}\right)^T\left(\mathbf{R}_z^{lat}\right)^{-1}\text{mat}\left(\underline{\xi}^{lat}\right)\mathbf{R}_h^{-1}\mathbf{\Phi}_h\right] \tag{5.15}$$

$\mathbf{L}^{lat}$ is the following lower triangular Cholesky decomposition:

$$\left(\mathbf{L}^{lat}\right)\left(\mathbf{L}^{lat}\right)^T=\sigma^2\mathbf{I}+\boldsymbol{\Omega}^{1/2}\times\left[\left(\boldsymbol{\Phi}_h^T\mathbf{R}_h^{-1}\boldsymbol{\Phi}_h\right)\otimes\left(\left(\boldsymbol{\Phi}_z^{lat}\right)^T\left(\mathbf{R}_z^{lat}\right)^{-1}\boldsymbol{\Phi}_z^{lat}\right)\right]\times\boldsymbol{\Omega}^{1/2} \tag{5.16}$$

$(L_{ii}^{lat}, L_{h,ii}, L_{z,ii}^{lat})$ are the ith diagonal terms of the lower triangular matrices $(\mathbf{L}^{lat}, \mathbf{L}_h, \mathbf{L}_z^{lat})$.

Regarding the term $\ln[f(\underline{\xi}^o|\underline{\theta}, \underline{\xi}^{lat})]$, note that $f(\underline{\xi}^o|\underline{\theta}, \underline{\xi}^{lat})$ is the marginal PDF of $f(\underline{\xi}^{\sim lat}|\underline{\theta}, \underline{\xi}^{lat})$ because $\underline{\xi}^o$ is a subset of $\underline{\xi}^{\sim lat}$. As shown in Appendix 5.1, $f(\underline{\xi}^{\sim lat}|\underline{\theta}, \underline{\xi}^{lat})$ is multivariate normal with the following mean vector $\underline{\mu}^{\sim lat|lat}\in\mathbf{R}^{(mo+mu)\times 1}$ and covariance matrix $\mathbf{C}^{\sim lat|lat}\in\mathbf{R}^{(mo+mu)\times(mo+mu)}$:

$$\begin{aligned}
\underline{\mu}^{\sim lat|lat}&=\mathrm{vec}\left[\boldsymbol{\Phi}_z^{\sim lat}\times\mathrm{mat}\left(\underline{\mu}^{\omega|lat}\right)\times\boldsymbol{\Phi}_h^T\right]\\
&\quad+\mathrm{vec}\left[\mathbf{R}_z^{\sim lat,lat}\left(\mathbf{R}_z^{lat}\right)^{-1}\mathrm{mat}\left(\underline{\xi}^{lat}-\mathrm{vec}\left[\boldsymbol{\Phi}_z^{lat}\times\mathrm{mat}\left(\underline{\mu}^{\omega|lat}\right)\times\boldsymbol{\Phi}_h^T\right]\right)\right]\\
\mathbf{C}^{\sim lat|lat}&=\sigma^2\left(\mathbf{R}_h\otimes\left[\mathbf{R}_z^{\sim lat}-\mathbf{R}_z^{\sim lat,lat}\left(\mathbf{R}_z^{lat}\right)^{-1}\mathbf{R}_z^{lat,\sim lat}\right]\right)\\
&\quad+\left(\boldsymbol{\Phi}_h\otimes\left[\boldsymbol{\Phi}_z^{\sim lat}-\mathbf{R}_z^{\sim lat,lat}\left(\mathbf{R}_z^{lat}\right)^{-1}\boldsymbol{\Phi}_z^{lat}\right]\right)\times\mathbf{C}^{\omega|lat}\\
&\quad\times\left(\boldsymbol{\Phi}_h\otimes\left[\boldsymbol{\Phi}_z^{\sim lat}-\mathbf{R}_z^{\sim lat,lat}\left(\mathbf{R}_z^{lat}\right)^{-1}\boldsymbol{\Phi}_z^{lat}\right]\right)^T
\end{aligned} \tag{5.17}$$

where

$$\begin{aligned}
\underline{\mu}^{\omega|lat}&=\sigma^{-2}\mathbf{C}^{\omega|lat}\times\mathrm{vec}\left[\left(\boldsymbol{\Phi}_z^{lat}\right)^T\left(\mathbf{R}_z^{lat}\right)^{-1}\times\mathrm{mat}\left(\underline{\xi}^{lat}\right)\times\mathbf{R}_h^{-1}\boldsymbol{\Phi}_h\right]\\
\mathbf{C}^{\omega|lat}&=\left[\boldsymbol{\Omega}^{-1}+\sigma^{-2}\left(\boldsymbol{\Phi}_h^T\mathbf{R}_h^{-1}\boldsymbol{\Phi}_h\right)\otimes\left(\left(\boldsymbol{\Phi}_z^{lat}\right)^T\left(\mathbf{R}_z^{lat}\right)^{-1}\boldsymbol{\Phi}_z^{lat}\right)\right]^{-1}
\end{aligned} \tag{5.18}$$

Note that Equations (5.17) and (5.18) do not require the inversion or Cholesky decomposition of large matrices, so the computation is fast. Because $\underline{\xi}^o$ is a subset of $\underline{\xi}^{\sim lat}$, $f(\underline{\xi}^o|\underline{\theta}, \underline{\xi}^{lat})$ is also multivariate normal with mean vector $\underline{\mu}^{o|lat}\in\mathbf{R}^{mo\times 1}$ being a sub-vector of $\underline{\mu}^{\sim lat|lat}$ and the covariance matrix $\mathbf{C}^{o|lat}\in\mathbf{R}^{mo\times mo}$ being a sub-matrix of $\mathbf{C}^{\sim lat|lat}$. As a result, $\ln[f(\underline{\xi}^o|\underline{\theta},\underline{\xi}^{lat})]$ can be computed by the following equation:

$$\begin{aligned}
\ln\left[f\left(\underline{\xi}^o\mid\underline{\theta},\underline{\xi}^{lat}\right)\right]&=-\frac{m^o}{2}\ln(2\pi)-\frac{1}{2}\ln\left(\left|\mathbf{C}^{o|lat}\right|\right)\\
&\quad-\frac{1}{2}\left(\underline{\xi}^o-\underline{\mu}^{o|lat}\right)^T\left(\mathbf{C}^{o|lat}\right)^{-1}\left(\underline{\xi}^o-\underline{\mu}^{o|lat}\right)
\end{aligned} \tag{5.19}$$

Equation (5.19) is prone to numerical round-off errors (e.g., $|\mathbf{C}^{ollat}|$ may be close to 0). The following equation is numerically robust:

$$\ln\left[f\left(\underline{\xi}^{o} \mid \underline{\theta}, \underline{\xi}^{lat}\right)\right] = -\frac{m^{o}}{2}\ln(2\pi) - \sum_{i=1}^{m^{o}} \ln\left(L_{ii}^{ollat}\right) - \frac{1}{2}\left(\underline{\alpha}^{ollat}\right)^{T}\left(\underline{\alpha}^{ollat}\right) \quad (5.20)$$

where $\mathbf{L}^{ollat}$ is the lower triangular Cholesky decomposition for $\mathbf{C}^{ollat}$, and $\underline{\alpha}^{ollat} = (\mathbf{L}^{ollat})^{-1}(\underline{\xi}^{o} - \underline{\mu}^{ollat})$. As mentioned earlier, the implementation of the GPR method requires the evaluation of the log-likelihood function $\ln[f(\underline{\xi}^{o},\underline{\xi}^{lat}|\underline{\theta})]$. Because $\ln[f(\underline{\xi}^{lat}|\underline{\theta})]$ can be evaluated efficiently using Equation (5.14) and $\ln[f(\underline{\xi}^{o}|\underline{\theta},\underline{\xi}^{lat})]$ can be evaluated efficiently using Equation (5.20), the computation for $\ln[f(\underline{\xi}^{o},\underline{\xi}^{lat}|\underline{\theta})]$ can be also efficient even though $\{\underline{\xi}^{o},\underline{\xi}^{lat}\}$ is non-lattice.

Given that the log-likelihood function $\ln[f(\underline{\xi}^{o},\underline{\xi}^{lat}|\underline{\theta})]$ can be evaluated efficiently, the TMCMC algorithm is adopted to draw posterior samples $\underline{\theta} \sim f(\underline{\theta}|\underline{\xi}^{o},\underline{\xi}^{lat})$. Non-informative (flat) prior PDFs are adopted for all unknown parameters in $\underline{\theta}$, including the auto-covariance parameters for the trend $\{\sigma^{(t)}, \delta_h^{(t)}, \delta_z^{(t)}\}$ and the auto-covariance parameters for the spatial variability $\{\sigma, \delta_h, \delta_z, \nu_h, \nu_z\}$. The sample size for each TMCMC stage is taken to be N = 2000. In the end of the TMCMC algorithm, 2000 samples of $\underline{\theta} \sim f(\underline{\theta}|\underline{\xi}^{o},\underline{\xi}^{lat})$ are obtained. Figure 5.7 shows how the posterior samples of $\underline{\theta}$. The scatter of these samples reflects the statistical uncertainty. From the figure, it is evident that (δ_z, ν_z) are identifiable ($\delta_z \approx 0.32$ m and $\nu_z \approx 1.4$), but (δ_h, ν_h) are not. In the following analysis, it is assumed that the horizontal smoothness is the same as the vertical smoothness ($\nu_h = \nu_z \approx 1.4$), whereas $\delta_h = 6$ m is assumed, which is similar to that reported by Stuedlein et al. (2012). Based on the $\{\delta_z, \delta_h, \nu_z, \nu_h\}$ values, $\mathbf{R}_h$ and $\mathbf{R}_z$ can be calculated. The trend parameters $\{\sigma^{(t)}, \delta_h^{(t)}, \delta_z^{(t)}\}$ are not adopted for the following analysis because it is assumed that the trend at the Baytown site can be captured by the cross-correlation. To elaborate this assumption, consider that a clay layer has s_u/σ'_v and q_{t1} with depth trends shown in Figure 5.8a and 5.8b. Although the data have clear depth trends, if the data at different depths are plotted in a s_u/σ'_v-q_{t1} plot (Figures 5.8c), a unique s_u/σ'_v-q_{t1} cross-correlation may exist. This suggests that the cross-correlation may model the depth trends. Because the cross-correlation of the Baytown site has been modeled by $(\underline{\mu}_s,\mathbf{C}_s)$, there is no need to model the depth trends separately.

5.5 BAYESIAN PARAMETER ESTIMATION FOR THE BAYTOWN SITE – ACCOMMODATE SPATIAL CORRELATION AND MISSING DATA

The purpose of this section is to estimate the quasi-site-specific mean vector $(\underline{\mu}_s)$ and covariance matrix $(\mathbf{C}_s)$ for the Baytown site. The HBM trained by the database $(\mathbf{X}^{o})$ can provide the prior model $f(\underline{\mu}_s,\mathbf{C}_s|\mathbf{X}^{o})$, and this prior PDF

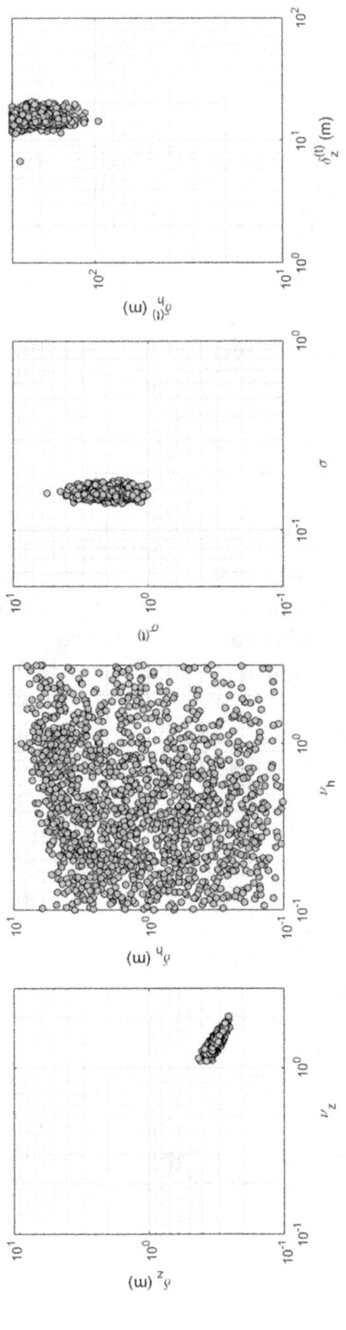

Figure 5.7 Illustration for non-lattice data.

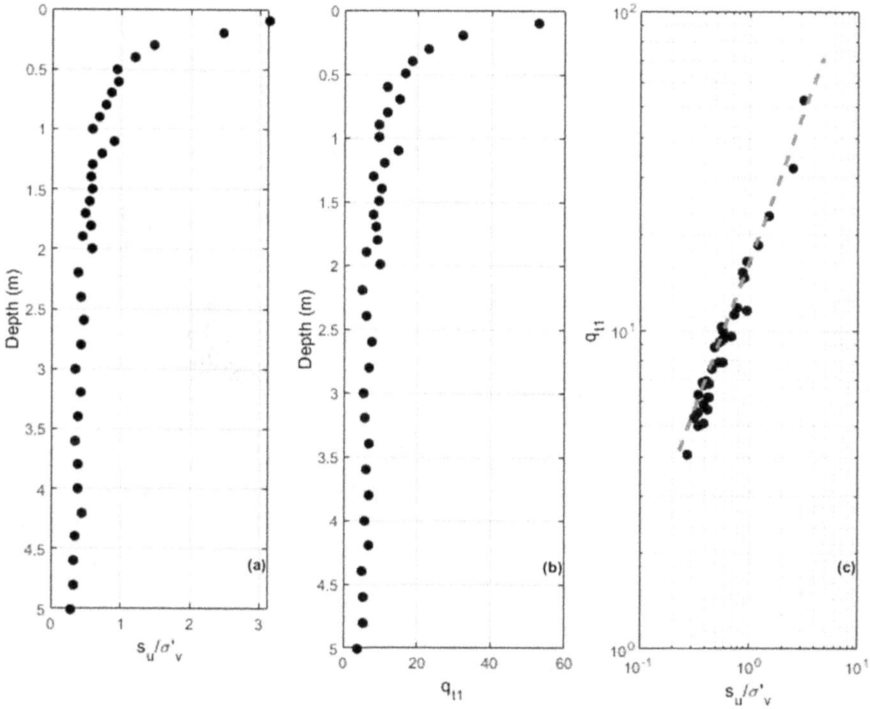

Figure 5.8 (a) Depth trend for s_u/σ'_v; (b) depth trend for q_{t1}; (c) s_u/σ'_v-q_{t1} relationship (Reprinted with permission from American Society of Civil Engineers (Ching and Phoon 2020)).

is further updated by the Baytown-site data ($\mathbf{X}_s^o$; the subscript "s" denotes "site-specific", and the superscript "o" denotes "observed") into the quasi-site-specific model $f(\underline{\mu}_s, C_s|\mathbf{X}^o, \mathbf{X}_s^o)$. According to Equation (4.34),

$$f(\underline{\mu}_s, C_s \mid \mathbf{X}^o, \mathbf{X}_s^o) \approx \frac{1}{N}\sum_{k=1}^{N} f(\underline{\mu}_s, C_s \mid \mathbf{\Theta}_k, \mathbf{X}_s^o) \tag{5.21}$$

where $\{\mathbf{\Theta}_k$: k = 1, ..., N$\}$ are the samples drawn from $f(\mathbf{\Theta}|\mathbf{X}^o)$. The samples $(\underline{\mu}_s, C_s) \sim f(\underline{\mu}_s, C_s|\mathbf{\Theta}, \mathbf{X}_s)$ can be drawn using a sub-GS procedure described in Section 4.1.3.4 if there is no spatial correlation among the target-site data $\mathbf{X}_s$ and if there is no missing data: $\underline{\mu}_s \sim f(\underline{\mu}_s|C_s, \mathbf{\Theta}, \mathbf{X}_s)$ using Equation (4.35) and $C_s \sim f(C_s|\underline{\mu}_s, \mathbf{\Theta}, \mathbf{X}_s)$ using Equation (4.36). However, there is spatial correlation and there are missing data for the Baytown-site data (e.g., Tables 5.1–5.3), so Equations (4.35) and (4.36) are no longer applicable. New equations that can accommodate the spatial correlation and missing data are needed. The new equations that can accommodate spatially correlated data are derived in Appendix 5.2, which shows that with the spatial correlation, $f(\underline{\mu}_s|C_s,\Theta,\underline{X}_s)$ is a multivariate normal PDF with the following mean vector $E(\underline{\mu}_s|C_s,\Theta,\underline{X}_s)$ and covariance matrix $Var(\underline{\mu}_s|C_s,\Theta,\underline{X}_s)$:

$$E(\underline{\mu}_s \mid C_s,\Theta,\underline{X}_s) = Var(\underline{\mu}_s \mid C_s,\Theta,\underline{X}_s) \cdot\left[C_0^{-1}\underline{\mu}_0 + C_s^{-1}mat(\underline{X}_s)\cdot vec\left(\mathbf{R}_z^{-1}\cdot\mathbf{1}_{m_z\times m_h}\cdot\mathbf{R}_h^{-1}\right)\right]$$

$$Var(\underline{\mu}_s \mid C_s,\Theta,\underline{X}_s) = \left[C_0^{-1} + \left(\underline{1}_{m_h\times 1}^T\mathbf{R}_h^{-1}\underline{1}_{m_h\times 1}\right)\left(\underline{1}_{m_z\times 1}^T\mathbf{R}_z^{-1}\underline{1}_{m_z\times 1}\right)C_s^{-1}\right]^{-1} \quad (5.22)$$

where $\mathbf{1}_{mz\times mh}$ is an $(m_z\times m_h)$ matrix containing ones; $\mathbf{R}_h$ and $\mathbf{R}_z$ denote the horizontal and vertical auto-correlation matrices for the spatial variability; $mat(\underline{X}_s)$ is an $(n\times m_s)$ matrix formed by stacking the column vectors $\underline{X}_{s,1}$, $\underline{X}_{s,2}$, ..., and $\underline{X}_{s,ms}$ horizontally; and vec(.) is the inverse process of mat(.): $vec(\mathbf{A}_{n\times ms})$ stacks the column vectors of matrix $\mathbf{A}$ into an $[(n\times m_s)\times 1]$ column vector. The auto-correlation matrices $\mathbf{R}_z$ and $\mathbf{R}_h$ of the Baytown site have been identified from the previous section. Appendix 5.2 also shows that with the spatial correlation, $f(C_s|\underline{\mu}_s,\Theta,\underline{X}_s)$ is an inverse-Wishart PDF with the degree of freedom = $\nu_0 + m_s$ and the scale matrix = $\mathbf{\Sigma}_0 + S$, where $S\in R^{n\times n}$ and its (i,j) entry S_{ij} can be computed as

$$S_{ij} = trace\left[mat\left(\underline{X}_{s,i.} - \mu_{s,i}\underline{1}_{m\times 1}^T\right)^T\cdot\mathbf{R}_z^{-1}\cdot mat\left(\underline{X}_{s,j.} - \mu_{s,j}\underline{1}_{m\times 1}^T\right)\cdot\mathbf{R}_h^{-1}\right] \quad (5.23)$$

where $\underline{X}_{s,i.}\in\mathbf{R}^{1\times m}$ is the ith row of $mat(\underline{X}_s)$, containing the ith soil property for all m locations; $\mu_{s,i}$ is the ith entry of $\underline{\mu}_s$, which is the mean value for the ith soil property.

The following procedure can be adopted to further update the prior PDF $f(\underline{\mu}_s,C_s|\mathbf{X}^o)$ into the quasi-site-specific model $f(\underline{\mu}_s,C_s|\mathbf{X}^o,\mathbf{X}_s^o)$. The procedure can also accommodate missing data of the Baytown site. Let us denote the collection of all missing data of the Baytown site by $\mathbf{X}_s^u$. Recall that the prior PDF $f(\underline{\mu}_s,C_s|\mathbf{X}^o)$ is captured by the hyper-parameter samples $\{\mathbf{\Theta}_k: i = 1, ..., N\}$ $= \{(\underline{\mu}_{0,k},C_{0,k},\mathbf{\Sigma}_{0,k},\nu_{0,k}): k = 1, ..., N\}$. For each set of hyper-parameter samples $(\underline{\mu}_{0,k},C_{0,k},\Sigma_{0,k},\nu_{0,k})$, do the following:

1. Initialize $(\underline{\mu}_s, C_s, \mathbf{X}_s^u)$ samples at arbitrary values. $\mathbf{X}_s$ is complete after the initiation.
2. Simulate $\underline{\mu}_s \sim f(\underline{\mu}_s|C_s,\mathbf{\Theta}_k,\underline{X}_s)$, a multivariate normal PDF with the following mean vector and covariance matrix:

$$E(\underline{\mu}_s \mid C_s,\Theta_k,\underline{X}_s) = Var(\underline{\mu}_s \mid C_s,\Theta_k,\underline{X}_s) \cdot\left[C_0^{-1}\underline{\mu}_{0,k} + C_s^{-1}mat(\underline{X}_s)\cdot vec\left(\mathbf{R}_z^{-1}\cdot\mathbf{1}_{m_z\times m_h}\cdot\mathbf{R}_h^{-1}\right)\right]$$

$$Var(\underline{\mu}_s \mid C_s,\Theta_k,\underline{X}_s) = \left[C_{0,k}^{-1} + \left(\underline{1}_{m_h\times 1}^T\mathbf{R}_h^{-1}\underline{1}_{m_h\times 1}\right)\left(\underline{1}_{m_z\times 1}^T\mathbf{R}_z^{-1}\underline{1}_{m_z\times 1}\right)C_s^{-1}\right]^{-1} \quad (5.24)$$

3. Simulate $\mathbf{C}_s \sim f(\mathbf{C}_s|\boldsymbol{\mu}_s,\boldsymbol{\Theta}_k,\mathbf{X}_s)$, an inverse-Wishart PDF with degree of freedom = $n + \nu_{0,k}$ and scale matrix equal to $\boldsymbol{\Sigma}_{0,k} + \mathbf{S}$, where

$$S_{ij} = \text{trace}\left[\text{mat}\left(\underline{X}_{s,i.} - \mu_{s,i}\underline{1}_{m\times 1}^{T}\right)^{T} \cdot \mathbf{R}_z^{-1} \cdot \text{mat}\left(\underline{X}_{s,j.} - \mu_{s,j}\underline{1}_{m\times 1}^{T}\right) \cdot \mathbf{R}_h^{-1}\right] \quad (5.25)$$

4. Simulate $\mathbf{X}_s^u \sim f(\mathbf{X}_s^u|\boldsymbol{\mu}_s,\mathbf{C}_s,\mathbf{X}_s^o)$, which is multivariate normal. This step can be computationally costly because of the need to simulate 3D data. In the current chapter, the sounding-wise GS method proposed by Yang and Ching (2021) is adopted to efficiently simulate $\mathbf{X}_s^u$. The sounding-wise GS is computationally feasible even with hundreds of soundings.
5. Repeat Steps 2–4 for T′ times to obtain T′ samples for ($\boldsymbol{\mu}_s$, $\mathbf{C}_s$). The ($\boldsymbol{\mu}_s$, $\mathbf{C}_s$) samples are collected after the burn-in period (T'_b). We simply take $T' = T'_b + 1$, so each hyper-parameter sample $\boldsymbol{\Theta}_k$ will produce one sample of ($\boldsymbol{\mu}_s$, $\mathbf{C}_s$, $\mathbf{X}_s^u$).

The above procedure is repeated N times for all hyper-parameter samples $\{\boldsymbol{\Theta}_k$: k = 1, …, N$\}$. The resulting N samples of ($\boldsymbol{\mu}_s$, $\mathbf{C}_s$, $\mathbf{X}_s^u$) are from the quasi-site-specific model $f(\boldsymbol{\mu}_s,\mathbf{C}_s,\mathbf{X}_s^u|\mathbf{X}^o,\mathbf{X}_s^o)$. These samples not only have absorbed the prior information extracted from $\mathbf{X}^o$ (e.g., CLAY/10/7490 database) but also have absorbed the information in $\mathbf{X}_s^o$ (e.g., Baytown-site data).

5.6 SIMULATION OF CONDITIONAL RANDOM FIELD – DEALING WITH INDEPENDENT SIMULATION DOMAIN

The purpose of this section is to simulate conditional random field (CRF) samples at unexplored locations at the Baytown site. Suppose that there are $m_{h'}$ x-y locations for the CRF simulation (called the simulation columns), and all columns share the same $m_{z'}$ simulation depths; thus, the simulation lattice consists of $m' = m_{h'} \times m_{z'}$ locations. Let us denote the soil properties in the simulation lattice by $\mathbf{X}'_s \in \mathbf{R}^{n\times m'}$. The main assumption for the CRF simulation method proposed in this section is that the simulation domain also follows a lattice structure. This condition can be easily achieved in practice by fitting the smallest lattice to the simulation domain. Note that there are two lattices thus far: (a) the lattice for the Baytown-site data (called the sounding lattice) – this is where the five boreholes and nine CPT soundings reside. There are m_h = 14 x-y locations and m_z = 98 depths; (b) the lattice for the simulation domain (called the simulation lattice) – this is where the CRF is to be simulated. Section 4.2.2.2 presented a CRF simulation method that can efficiently simulate CRF with the restriction that the simulation domain shares the same depths as the sounding domain. In this section, a more general CRF simulation method without this restriction is proposed: the simulation lattice can be completely independent of the sounding lattice.

The purpose of the CRF simulation is to simulate $\mathbf{X}'_s \sim f(\mathbf{X}'_s|\mathbf{X}^o_s, \mathbf{X}^o)$. According to the total probability theory,

$$f(\mathbf{X}'_s \mid \mathbf{X}^o_s, \mathbf{X}^o) = \int f(\mathbf{X}'_s \mid \underline{\mu}_s, \mathbf{C}_s, \mathbf{X}^u_s, \mathbf{X}^o_s, \mathbf{X}^o) f(\underline{\mu}_s, \mathbf{C}_s, \mathbf{X}^u_s \mid \mathbf{X}^o_s, \mathbf{X}^o) d\underline{\mu}_s d\mathbf{C}_s d\mathbf{X}^u_s$$

$$\approx \frac{1}{N}\left[\sum_{k=1}^{N} f(\mathbf{X}'_s \mid \underline{\mu}_{s,k}, \mathbf{C}_{s,k}, \mathbf{X}^u_{s,k}, \mathbf{X}^o_s, \mathbf{X}^o)\right] = \frac{1}{N}\left[\sum_{k=1}^{N} f(\mathbf{X}'_s \mid \underline{\mu}_{s,k}, \mathbf{C}_{s,k}, \mathbf{X}^u_{s,k}, \mathbf{X}^o_s)\right] \tag{5.26}$$

where $(\underline{\mu}_{s,k}, \mathbf{C}_{s,k}, \mathbf{X}_{s,k}{}^u) \sim f(\underline{\mu}_s, \mathbf{C}_s, \mathbf{X}_s{}^u|\mathbf{X}^o, \mathbf{X}_s{}^o)$ is the kth sample obtained from the previous section. For each $(\underline{\mu}_{s,k}, \mathbf{C}_{s,k}, \mathbf{X}_{s,k}{}^u)$ sample, a sample of $\mathbf{X}'_s$ is drawn from $f(\mathbf{X}'_s|\underline{\mu}_{s,k}, \mathbf{C}_{s,k}, \mathbf{X}_{s,k}{}^u, \mathbf{X}^o{}_s)$. By doing so for all $\{(\underline{\mu}_{s,k}, \mathbf{C}_{s,k}, \mathbf{X}^u{}_{s,k}$: $k = 1, \ldots, N)$, N samples of $\mathbf{X}'_s$ are obtained, and, according to Equation (5.26), these samples are distributed as $f(\mathbf{X}'_s|\mathbf{X}^o{}_s, \mathbf{X}^o)$. The key step above is to simulate $\mathbf{X}'_s \sim f(\mathbf{X}'_s|\underline{\mu}_s, \mathbf{C}_s, \mathbf{X}_s)$, and the detailed steps for this simulation are described below.

Let us first define the following five lattices. Two of them (#1 and #5) were already defined and three of them (#2, #3, and #4) are new. For clarity, the non-primed subscript h denotes the x-y locations for the sounding lattice, whereas the non-primed subscript z denotes the depths for the sounding lattice. In contrast, the primed subscript h′ denotes the x-y locations for the simulation lattice, whereas the primed subscript z′ denotes the depths for the simulation lattice:

1. $\mathbf{X}_{hz} \in \mathbf{R}^{n \times m}$ is the lattice that adopts the m_h sounding x-y locations and the m_z sounding depths. This lattice is exactly the sounding lattice $\mathbf{X}_s$.
2. $\mathbf{X}_{hz'} \in \mathbf{R}^{n \times (mh \times mz')}$ is the hybrid lattice that adopts the m_h sounding x-y locations and the $m_{z'}$ simulation depths.
3. $\mathbf{X}_{h(z \cup z')} \in \mathbf{R}^{n \times [mh \times (mz+mz')]}$ is the lattice that adopts the m_h sounding x-y locations and the $m_z + m_{z'}$ union depths ($z \cup z'$ denotes the union between the sounding and simulation depths).
4. $\mathbf{X}_{h'(z \cup z')} \in \mathbf{R}^{n \times [mh' \times (mz+mz')]}$ is the lattice that adopts the $m_{h'}$ simulation columns and the $m_z + m_{z'}$ union depths.
5. $\mathbf{X}_{h'z'} \in \mathbf{R}^{n \times m'}$ is the lattice that adopts the $m_{h'}$ simulation columns and the $m_{z'}$ simulation depths. This lattice is exactly the simulation lattice $\mathbf{X}'_s$.

Let us also define the following matrices:

1. $\mathbf{R}_{h'}$ is the auto-correlation matrix for the x-y simulation locations. $\mathbf{R}_{z'}$ is the auto-correlation matrix for the simulation depths. $\mathbf{R}_{z \cup z'}$ is the auto-correlation matrix for the union depths ($z \cup z'$).
2. $\mathbf{R}_{h'h}$ is the cross-correlation matrix between the simulation and sounding x-y locations. $\mathbf{R}_{z'z}$ is the cross-correlation matrix between the simulation depths and the sounding depths.
3. The derivations provided in Appendix 5.3 demonstrates that $\mathbf{R}_{h'|h}$ is *defined* as $(\mathbf{R}_{h'} - \mathbf{R}_{h'h} \times \mathbf{R}_h^{-1} \times \mathbf{R}_{hh'})$, and $\mathbf{R}_{z'|z}$ is *defined* as $(\mathbf{R}_{z'} - \mathbf{R}_{z'z} \times \mathbf{R}_z^{-1} \times \mathbf{R}_{zz'})$.

4. The symbol **L** stands for the lower Choslesky decomposition. For instance, $\mathbf{L}_s$ is the lower Cholesky decomposition of $\mathbf{C}_s$, $\mathbf{L}_h$ is the lower Cholesky decomposition of $\mathbf{R}_h$, $\mathbf{L}_{h'|h}$ is the lower Cholesky decomposition of $\mathbf{R}_{h'|h}$, etc.

The simulation of $\mathbf{X}'_s \sim f(\mathbf{X}'_s|\underline{\mu}_s, \mathbf{C}_s, \mathbf{X}_s)$ consists of four steps:

1. Simulate $\mathbf{X}_{hz'} \sim f(\mathbf{X}_{hz'}|\underline{\mu}_s, \mathbf{C}_s, \mathbf{X}_{hz})$ ($\mathbf{X}_{hz}$ is the same as $\mathbf{X}_s$):

$$\mathbf{X}_{hz'} = \underline{\mu}_s \times \underline{\mathbf{1}}^T_{(m_h \times m_{z'})\times 1} + \left(\mathbf{X}_{hz} - \underline{\mu}_s \times \underline{\mathbf{1}}^T_{(m_h \times m_z)\times 1}\right)\left(\mathbf{I}_{m_h \times m_h} \otimes \mathbf{R}_z^{-T}\mathbf{R}_{z'z}^T\right)$$
$$+ \mathbf{L}_s\mathbf{Z}_{n\times(m_h \times m_{z'})}\left(\mathbf{L}_h^T \otimes \mathbf{L}_{z'|z}^T\right) \tag{5.27}$$

where $\mathbf{Z}_{n\times(mh\times mz')} \in \mathbf{R}^{n\times[mh\times(mz+mz')]}$ is a matrix that contains independent standard normal variables. $\mathbf{X}_{hz'}$ can be simulated by a simple equation (Equation 5.27) because $\mathbf{X}_{hz'}$ and $\mathbf{X}_{hz}$ share the same horizontal locations (both adopt subscript h).
2. Combine $\mathbf{X}_{hz'}$ and $\mathbf{X}_{hz}$ into $\mathbf{X}_{h(z\cup z')}$.
3. Simulate $\mathbf{X}_{h'(z\cup z')} \sim f(\mathbf{X}_{h'(z\cup z')}|\underline{\mu}_s, \mathbf{C}_s, \mathbf{X}_{h(z\cup z')})$.

$$\mathbf{X}_{h'(z\cup z')} = \underline{\mu}_s \times \underline{\mathbf{1}}^T_{[m_{h'}\times(m_z+m_{z'})]\times 1} + \left(\mathbf{X}_{h(z\cup z')} - \underline{\mu}_s \times \underline{\mathbf{1}}^T_{[m_{h'}\times(m_z+m_{z'})]\times 1}\right)$$
$$\left(\mathbf{R}_h^{-T}\mathbf{R}_{h'h}^T \otimes \mathbf{I}_{(m_z+m_{z'})\times(m_z+m_{z'})}\right) + \mathbf{L}_s\mathbf{Z}_{n\times[m_{h'}\times(m_z+m_{z'})]}\left(\mathbf{L}_{h'|h}^T \otimes \mathbf{L}_{z\cup z'}^T\right) \tag{5.28}$$

$\mathbf{X}_{h'(z\cup z')}$ can be simulated by a simple equation (Equation 5.28) because $\mathbf{X}_{h'(z\cup z')}$ and $\mathbf{X}_{h(z\cup z')}$ share the same depths (both adopt subscript $z\cup z'$).
4. Extract $\mathbf{X}_{h'z'}$ from $\mathbf{X}_{h'(z\cup z')}$ (because $\mathbf{X}_{h'z'}$ is a subset of $\mathbf{X}_{h'(z\cup z')}$). Note that $\mathbf{X}'_s$ is exactly $\mathbf{X}_{h'z'}$.

The derivations for Equations (5.27) and (5.28) are detailed in Appendix 5.3. Note that there are fast algorithms (e.g., Fernandes et al. 1998) to compute **A**×(**B**⊗**C**) in Equations (5.27) and (5.28).

To illustrate the CRF simulation results, consider the simulation lattice defined by 31 x-grid points = (0 m, 0.5 m, …, 14.5 m, 15 m), 61 y-grid points = (0 m, 0.5 m, …, 29.5 m, 30 m), and 181 z-grid points = (1 m, 1.05 m, …, 9.95 m, 10 m). Therefore, there are $m_{z'} = 181$ simulation depths and $m_{h'} = 31\times61 = 1891$ simulation locations. The proposed method can simulate the CRF samples over the simulation lattice for any soil parameters, for example, Figure 5.9 shows one realization for the CRF of the undrained shear strength (s_u) projected onto three slice sections. To further illustrate the simulation of the CRF in detail, consider the three simulation locations (#1–#3) in particular:

1. The #1 location coincides with the x-y location of B-3 in Figure 5.1.
2. The #2 location coincides with the x-y location of CPT-1 in Figure 5.1.
3. The #3 location is 1 m south of B-3.

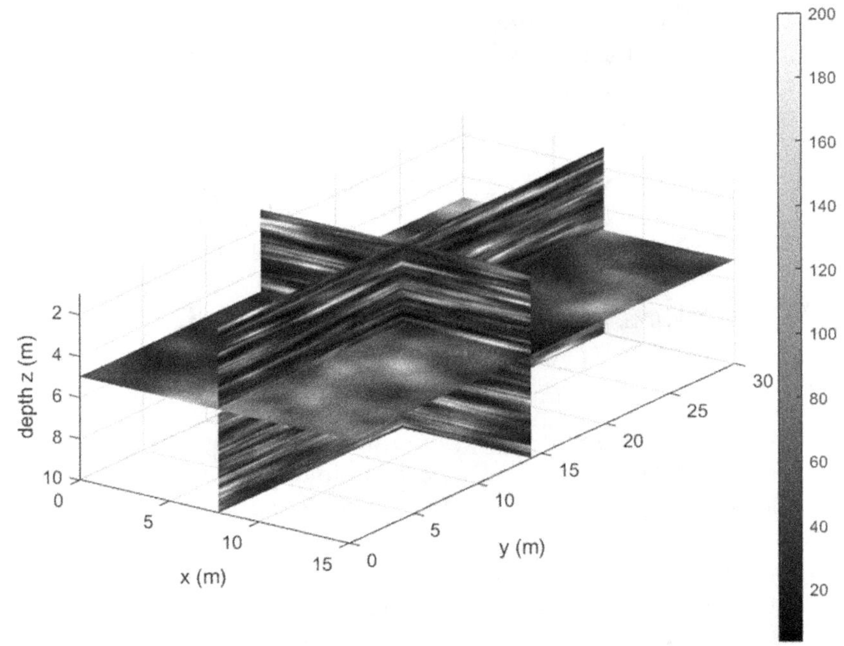

Figure 5.9 Realization of s_u CRF (projected onto three slice sections) (s_u value in kPa).

Figure 5.10 shows the simulation results for the #1 location (coincides with B-3), whereas Figure 5.11 shows the results for the #2 location (coincides with CPT-1). The grey solid line shows one realization of the CRF, the dark solid line shows the median of the CRF, and the dark dashed lines show the 95% confidence interval (CI) of the CRF. The actual data points for B-3 are also shown as circles for comparison. It is clear that the 95% CI shrinks to zero width whenever the simulation depth coincides with the measurement depth. The gray patch zone in the figure indicates the depth range [3.4 m, 4.7 m] for the silty sand layer: this depth range is not analyzed. Based on the Baytown data, Stuedlein et al. (2012) estimated a trend of OCR with depth, shown as the thick solid line in Figures 5.10d and 5.11d. The originally proposed trend of OCR is generally consistent with the prediction results (median and 95% CI). Also based on the Baytown data, Stuedlein et al. (2012) proposed an SHANSEP equation that can be used to estimate s_u based on LL and OCR. The thick solid lines in Figures 5.10e and 5.11e show the estimated mean trend of s_u based on the mean trends of LL and OCR. The mean trend of s_u is consistent with the median value of the s_u CRF. Figure 5.12 shows the simulation results for the #3 location (1 m south of B-3). The purpose of Figure 5.12 is to illustrate the CRFs at an x-y location where a borehole/sounding is not conducted. Unlike the #1 and #2 locations that coincide with B-3 and CPT-1, the #3 location is an un-sampled location. The results (median and 95% CI) are similar to those for the #1 location (Figure 5.10) because the separation distance between the #1 and #3 location is 1 m, which is only 1/6 of the assumed δ_h.

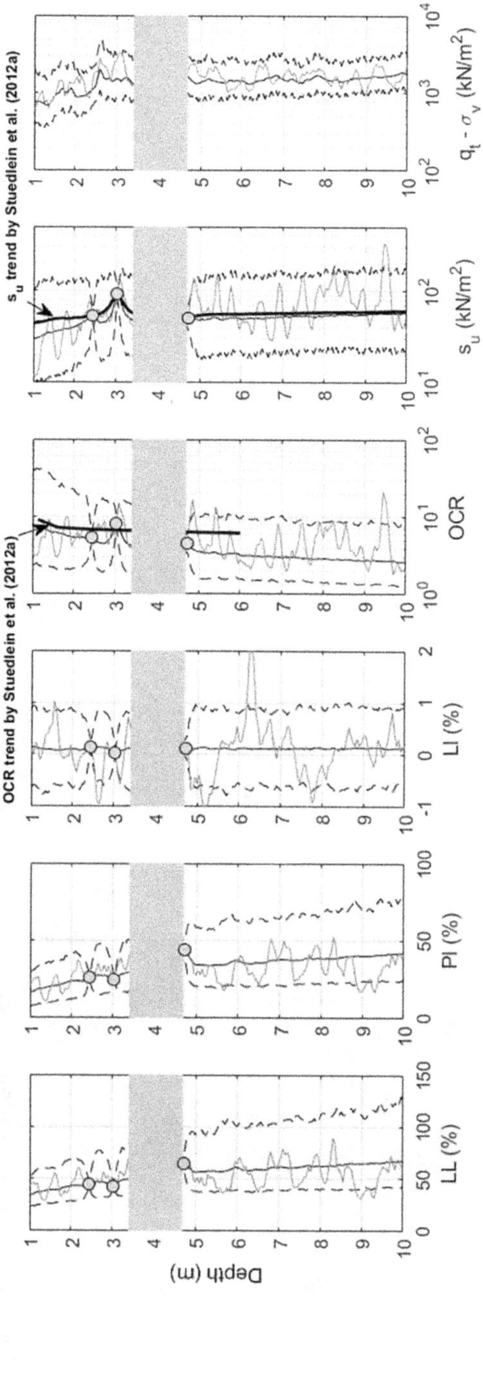

Figure 5.10 CRF simulation results for #1 location (coincides with B-3).

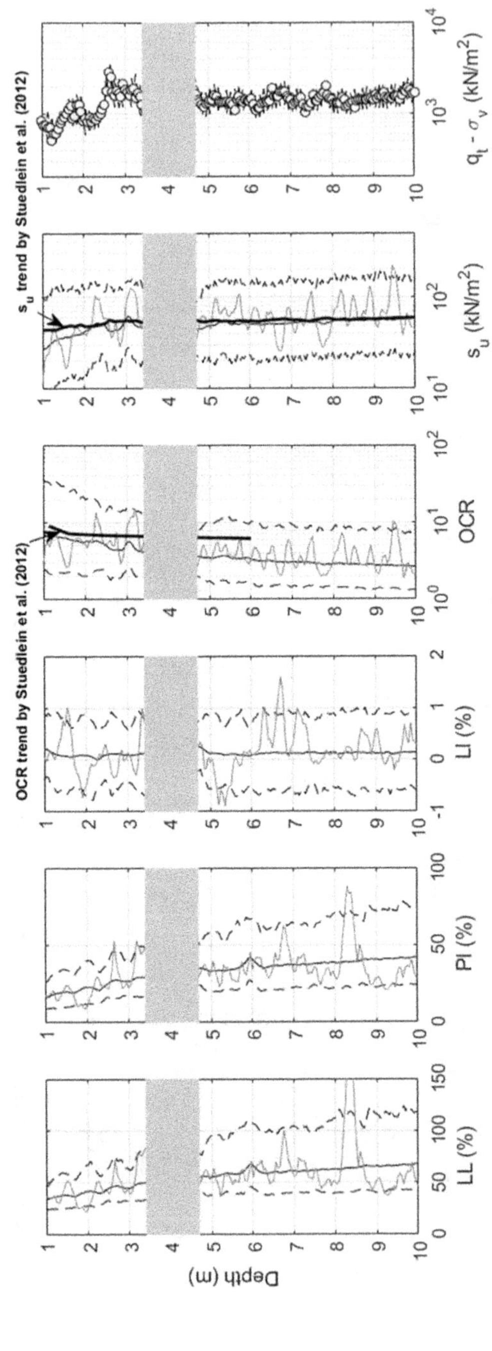

Figure 5.11 CRF simulation results for #2 location (coincides with CPT-1).

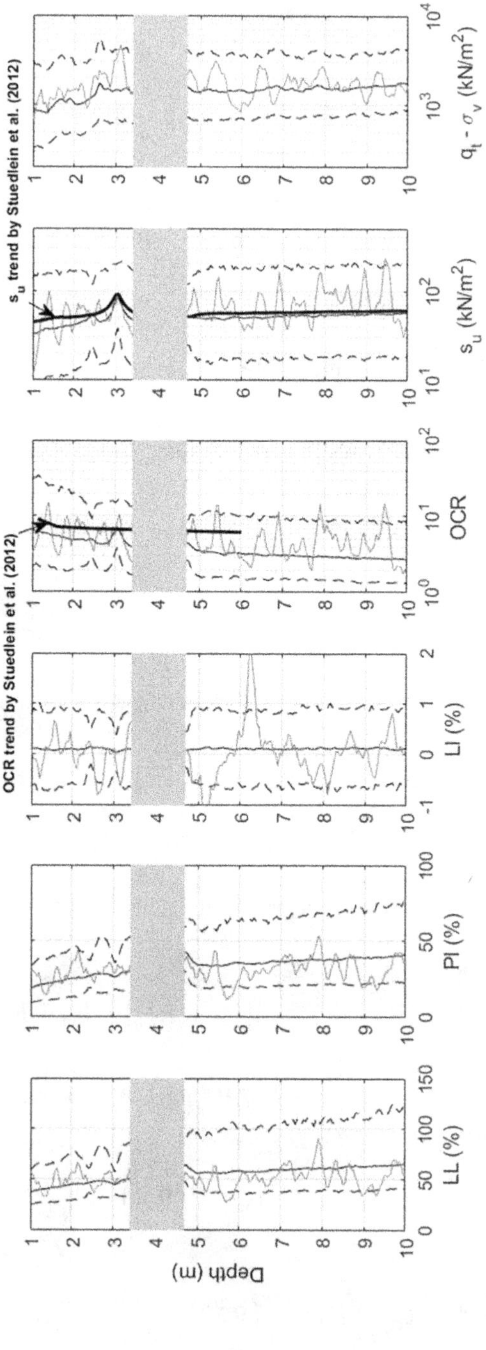

Figure 5.12 CRF simulation results for #3 location (1 m south of B-3).

APPENDIX 5.1: DERIVATIONS FOR THE MEAN VECTOR AND COVARIANCE MATRIX FOR $f(\underline{\xi}^{\sim lat}|\underline{\theta}, \underline{\xi}^{lat})$

Recall that $\underline{\xi}^{lat}$ can be expressed as

$$\underline{\xi}^{lat} = \left(\Phi_h \otimes \Phi_z^{lat}\right) \times \underline{\omega} + \underline{w}^{lat} \tag{A.19}$$

where $\underline{\omega} \in \mathbf{R}^{pq\times 1}$ is a zero-mean normal random vector with covariance matrix = $\mathbf{\Omega}$; $\underline{w}^{lat} \in \mathbf{R}^{(m_z lat \times m_h)\times 1}$ contains the spatial variabilities over the lattice depths. Following the similar derivations in Equations (A.15) and (A.16), $f(\underline{\omega}|\underline{\theta},\underline{\xi}^{lat})$ is multivariate normal with the mean vector $\underline{\mu}^{\omega|lat} \in \mathbf{R}^{pq\times 1}$ and covariance matrix $\mathbf{C}^{\omega|lat} \in \mathbf{R}^{pq\times pq}$:

$$\begin{aligned} \underline{\mu}^{\omega|lat} &= \sigma^{-2}\mathbf{C}^{\omega|lat} \times \mathrm{vec}\left[\left(\Phi_z^{lat}\right)^T \left(\mathbf{R}_z^{lat}\right)^{-1} \times \mathrm{mat}\left(\underline{\xi}^{lat}\right) \times \mathbf{R}_h^{-1}\Phi_h\right] \\ \mathbf{C}^{\omega|lat} &= \left[\Omega^{-1} + \sigma^{-2}\left(\Phi_h^T \mathbf{R}_h^{-1}\Phi_h\right) \otimes \left(\left(\Phi_z^{lat}\right)^T \left(\mathbf{R}_z^{lat}\right)^{-1} \Phi_z^{lat}\right)\right]^{-1} \end{aligned} \tag{A.20}$$

Because $\underline{t} = (\mathbf{\Phi}_h \otimes \mathbf{\Phi}_z) \times \underline{\omega}$, the mean vector and the covariance matrix for $f(\underline{t}|\underline{\theta},\underline{\xi}^{lat})$ are

$$\underline{\mu}^{t|lat} = E(\underline{t} \mid \theta, \underline{\xi}^{lat}) = \left(\Phi_h \otimes \Phi_z\right) \times \underline{\mu}^{\omega|lat} \tag{A.21}$$

$$\mathbf{C}^{t|lat} = \mathrm{Var}(\underline{t} \mid \theta, \underline{\xi}^{lat}) = \left(\Phi_h \otimes \Phi_z\right) \times \mathbf{C}^{\omega|lat} \times \left(\Phi_h \otimes \Phi_z\right)^T \tag{A.22}$$

where $\underline{\mu}^{t|lat}$ and $\mathbf{C}^{t|lat}$ denote the mean vector and the covariance matrix for $f(\underline{t}|\underline{\theta},\underline{\xi}^{lat})$, respectively. On the other hand, the mean vector and the covariance matrix for $f(\underline{\xi}^{\sim lat}|\underline{\theta},\underline{t},\underline{\xi}^{lat})$ can be written as

$$\begin{aligned} &E(\underline{\xi}^{\sim lat} \mid \underline{\theta}, \underline{t}, \underline{\xi}^{lat}) = \underline{t}^{\sim lat} + \left(\mathbf{R}_h \otimes \mathbf{R}_z^{\sim lat,lat}\right)\left(\mathbf{R}_h \otimes \mathbf{R}_z^{lat}\right)^{-1}\left(\underline{\xi}^{lat} - \underline{t}^{lat}\right) \\ &\mathrm{Var}(\underline{\xi}^{\sim lat} \mid \underline{\theta}, \underline{t}, \underline{\xi}^{lat}) \\ &= \sigma^2\left[\left(\mathbf{R}_h \otimes \mathbf{R}_z^{\sim lat}\right) - \left(\mathbf{R}_h \otimes \mathbf{R}_z^{\sim lat,lat}\right)\left(\mathbf{R}_h \otimes \mathbf{R}_z^{lat}\right)^{-1}\left(\mathbf{R}_h \otimes \mathbf{R}_z^{\sim lat,lat}\right)^T\right] \end{aligned} \tag{A.23}$$

Given $(\mathbf{A}\otimes\mathbf{B})^{-1} = \mathbf{A}^{-1}\otimes\mathbf{B}^{-1}$, $(\mathbf{A}\otimes\mathbf{B})^T = \mathbf{A}^T\otimes\mathbf{B}^T$, and $(\mathbf{A}^T\otimes\mathbf{B})\times\mathrm{vec}(\mathbf{C}) = \mathrm{vec}(\mathbf{BCA})$, Equation (5.23) becomes

$$\begin{aligned} &E(\underline{\xi}^{\sim lat} \mid \underline{\theta}, \underline{t}, \underline{\xi}^{lat}) = \underline{t}^{\sim lat} + \left(\mathbf{I}_{m_h \times m_h} \otimes \left[\mathbf{R}_z^{\sim lat,lat}\left(\mathbf{R}_z^{lat}\right)^{-1}\right]\right)\left(\underline{\xi}^{lat} - \underline{t}^{lat}\right) \\ &\mathrm{Var}(\underline{\xi}^{\sim lat} \mid \underline{\theta}, \underline{t}, \underline{\xi}^{lat}) \\ &= \sigma^2\left(\mathbf{R}_h \otimes \left[\mathbf{R}_z^{\sim lat} - \mathbf{R}_z^{\sim lat,lat}\left(\mathbf{R}_z^{lat}\right)^{-1}\left(\mathbf{R}_z^{\sim lat,lat}\right)^T\right]\right) = \sigma^2\left(\mathbf{R}_h \otimes \mathbf{R}_{z|lat}^{\sim lat}\right) \end{aligned} \tag{A.24}$$

where

$$\mathbf{R}_{z|lat}^{\sim lat} = \mathbf{R}_{z}^{\sim lat} - \mathbf{R}_{z}^{\sim lat,lat}\left(\mathbf{R}_{z}^{lat}\right)^{-1}\mathbf{R}_{z}^{lat,\sim lat} \tag{A.25}$$

Given the identities $E(X) = E_Y(E_X(X|Y))$, we have

$$\begin{aligned} \underline{\mu}^{\sim lat|lat} &= E(\underline{\xi}^{\sim lat} \mid \underline{\theta}, \underline{\xi}^{lat}) = E_{t|\theta,\underline{\xi}^{lat}}\left[E(\underline{\xi}^{\sim lat} \mid \underline{\theta}, \underline{t}, \underline{\xi}^{lat})\right] \\ &= E_{t|\theta,\underline{\xi}^{lat}}\left(\underline{t}^{\sim lat}\right) + \mathrm{vec}\left(\mathbf{R}_{z}^{\sim lat,lat}\left(\mathbf{R}_{z}^{lat}\right)^{-1}\mathrm{mat}\left[\underline{\xi}^{lat} - E_{t|\theta,\underline{\xi}^{lat}}\left(\underline{t}^{lat}\right)\right]\right) \\ &= \underline{\mu}^{t^{\sim lat}|lat} + \mathrm{vec}\left[\mathbf{R}_{z}^{\sim lat,lat}\left(\mathbf{R}_{z}^{lat}\right)^{-1}\mathrm{mat}\left(\underline{\xi}^{lat} - \underline{\mu}^{t^{lat}|lat}\right)\right] \end{aligned} \tag{A.26}$$

where $\underline{\mu}^{t^{lat}|lat}$ and $\underline{\mu}^{t^{\sim lat}|lat}$ are sub-vectors of the $\underline{\mu}^{t|lat}$ vector in Equation (5.21):

$$\begin{aligned} \underline{\mu}^{t^{lat}|lat} &= \left(\Phi_h \otimes \Phi_z^{lat}\right) \times \underline{\mu}^{\omega|lat} = \mathrm{vec}\left(\Phi_z^{lat} \times \mathrm{mat}\left(\underline{\mu}^{\omega|lat}\right) \times \Phi_h^{T}\right) \\ \underline{\mu}^{t^{\sim lat}|lat} &= \left(\Phi_h \otimes \Phi_z^{\sim lat}\right) \times \underline{\mu}^{\omega|lat} = \mathrm{vec}\left(\Phi_z^{\sim lat} \times \mathrm{mat}\left(\underline{\mu}^{\omega|lat}\right) \times \Phi_h^{T}\right) \end{aligned} \tag{A.27}$$

Given the identities $\mathrm{Var}(X) = E_Y(\mathrm{Var}_X(X|Y)) + \mathrm{Var}_Y(E_X(X|Y))$, we have

$$\begin{aligned} \mathbf{C}^{\sim lat|lat} &= \mathrm{Var}(\underline{\xi}^{\sim lat} \mid \underline{\theta}, \underline{\xi}^{lat}) = E_{t|\theta,\underline{\xi}^{lat}}\left[\mathrm{Var}(\underline{\xi}^{\sim lat} \mid \underline{\theta}, \underline{t}, \underline{\xi}^{lat})\right] \\ &\quad + \mathrm{Var}_{t|\theta,\underline{\xi}^{lat}}\left[E(\underline{\xi}^{\sim lat} \mid \underline{\theta}, \underline{t}, \underline{\xi}^{lat})\right] \\ &= \sigma^2\left(\mathbf{R}_h \otimes \mathbf{R}_{z|lat}^{\sim lat}\right) + \mathrm{Var}_{t|\theta,\underline{\xi}^{lat}}\left[\underline{t}^{\sim lat} + \left(\mathbf{I}_{m_h \times m_h} \otimes \left[\mathbf{R}_{z}^{\sim lat,lat}\left(\mathbf{R}_{z}^{lat,lat}\right)^{-1}\right]\right)\left(\underline{\xi}^{lat} - \underline{t}^{lat}\right)\right] \\ &= \sigma^2\left(\mathbf{R}_h \otimes \mathbf{R}_{z|lat}^{\sim lat}\right) + \mathbf{C}^{t^{\sim lat}|lat} + \left(\mathbf{I}_{m_h \times m_h} \otimes \left[\mathbf{R}_{z}^{\sim lat,lat}\left(\mathbf{R}_{z}^{lat}\right)^{-1}\right]\right) \\ &\quad \mathbf{C}^{t^{lat}|lat}\left(\mathbf{I}_{m_h \times m_h} \otimes \left[\mathbf{R}_{z}^{\sim lat,lat}\left(\mathbf{R}_{z}^{lat}\right)^{-1}\right]\right)^{T} - 2\left(\mathbf{I}_{m_h \times m_h} \otimes \left[\mathbf{R}_{z}^{\sim lat,lat}\left(\mathbf{R}_{z}^{lat}\right)^{-1}\right]\right)\mathbf{C}^{t^{lat},t^{\sim lat}|lat} \end{aligned} \tag{A.28}$$

where $\mathbf{C}^{t^{lat}|lat}$, $\mathbf{C}^{t^{\sim lat}|lat}$, and $\mathbf{C}^{t^{lat},t^{\sim lat}|lat}$ are sub-matrices of the $\mathbf{C}^{t|lat}$ matrix in Equation (5.22):

$$\begin{aligned} \mathbf{C}^{t^{lat}|lat} &= \left(\Phi_h \otimes \Phi_z^{lat}\right) \times \mathbf{C}^{\omega|lat} \times \left(\Phi_h \otimes \Phi_z^{lat}\right)^{T} \\ \mathbf{C}^{t^{\sim lat}|lat} &= \left(\Phi_h \otimes \Phi_z^{\sim lat}\right) \times \mathbf{C}^{\omega|lat} \times \left(\Phi_h \otimes \Phi_z^{\sim lat}\right)^{T} \\ \mathbf{C}^{t^{lat},t^{\sim lat}|lat} &= \left(\Phi_h \otimes \Phi_z^{lat}\right) \times \mathbf{C}^{\omega|lat} \times \left(\Phi_h \otimes \Phi_z^{\sim lat}\right)^{T} \end{aligned} \tag{A.29}$$

Inserting Equation (A.29) into Equation (5.28) and applying the identity $(\mathbf{AB})\otimes(\mathbf{CD}) = (\mathbf{A}\otimes\mathbf{C})(\mathbf{B}\otimes\mathbf{D})$ yields

$$\mathbf{C}^{\sim lat|lat} = \sigma^2\left(\mathbf{R}_h \otimes \mathbf{R}_z^{\sim lat|lat}\right) + \left(\boldsymbol{\Phi}_h \otimes \boldsymbol{\Phi}_z^{\sim lat|lat}\right) \times \mathbf{C}^{\omega|lat} \times \left(\boldsymbol{\Phi}_h \otimes \boldsymbol{\Phi}_z^{\sim lat|lat}\right)^T \quad \text{(A.30)}$$

where

$$\boldsymbol{\Phi}_z^{\sim lat|lat} = \boldsymbol{\Phi}_z^{\sim lat} - \mathbf{R}_z^{\sim lat,lat}\left(\mathbf{R}_z^{lat}\right)^{-1}\boldsymbol{\Phi}_z^{lat} \quad \text{(A.31)}$$

Equations (5.26) and (A.30) are identical to Equation (5.17).

APPENDIX 5.2: DERIVATIONS FOR $f(\underline{\mu}_s|\mathbf{C}_s,\boldsymbol{\Theta},\mathbf{X}_s)$ AND $f(\mathbf{C}_s|\underline{\mu}_s,\boldsymbol{\Theta},\mathbf{X}_s)$ UNDER SPATIAL CORRELATION

This appendix derives the analytical forms for the full conditional PDFs $f(\underline{\mu}_s|\mathbf{C}_s,\boldsymbol{\Theta},\mathbf{X}_s)$ and $f(\mathbf{C}_s|\underline{\mu}_s,\boldsymbol{\Theta},\mathbf{X}_s)$ when the spatial correlation is present. The derivations start from the expression of the complete multivariate PDF of $(\underline{\mu}_s, \mathbf{C}_s, \boldsymbol{\Theta}, \mathbf{X}_s)$:

$$\begin{aligned} f\left(\underline{\mu}_s,\mathbf{C}_s,\boldsymbol{\Theta},\mathbf{X}_s\right) &= f(\underline{\mathbf{X}}_s \mid \underline{\mu}_s,\mathbf{C}_s)\cdot f(\underline{\mu}_s \mid \underline{\mu}_0,\mathbf{C}_0)\cdot f(\mathbf{C}_s \mid \Sigma_0,\nu_0) \\ &\propto \left|\mathbf{R}\otimes\mathbf{C}_s\right|^{-\frac{1}{2}}\cdot e^{-\frac{1}{2}\left(\underline{\mathbf{X}}_s-\underline{1}_{m\times1}\otimes\underline{\mu}_s\right)^T(\mathbf{R}\otimes\mathbf{C}_s)^{-1}\left(\underline{\mathbf{X}}_s-\underline{1}_{m\times1}\otimes\underline{\mu}_s\right)} \\ &\times\left|\mathbf{C}_0\right|^{-\frac{1}{2}}\cdot e^{-\frac{1}{2}\left(\underline{\mu}_s-\underline{\mu}_0\right)^T\mathbf{C}_0^{-1}\left(\underline{\mu}_s-\underline{\mu}_0\right)}\times\frac{\left|\Sigma_0\right|^{\nu_0/2}\cdot\left|\mathbf{C}_s\right|^{-\frac{\nu_0+n+1}{2}}}{2^{n\times\nu_0/2}\cdot\Gamma_n\left(\nu_0/2\right)}\cdot e^{-\frac{1}{2}tr\left(\Sigma_0\times\mathbf{C}_s^{-1}\right)} \end{aligned} \quad \text{(A.32)}$$

For the full conditional PDF $f(\underline{\mu}_s|\mathbf{C}_s,\boldsymbol{\Theta},\mathbf{X}_s)$, one has to focus on the $\underline{\mu}_s$-related terms in Equation (A.32):

$$f(\underline{\mu}_s \mid \mathbf{C}_s,\boldsymbol{\Theta},\mathbf{X}_s) \propto e^{-\frac{1}{2}\left(\underline{\mathbf{X}}_s-\underline{1}_{m\times1}\otimes\underline{\mu}_s\right)^T(\mathbf{R}\otimes\mathbf{C}_s)^{-1}\left(\underline{\mathbf{X}}_s-\underline{1}_{m\times1}\otimes\underline{\mu}_s\right)-\frac{1}{2}\left(\underline{\mu}_s-\underline{\mu}_0\right)^T\mathbf{C}_0^{-1}\left(\underline{\mu}_s-\underline{\mu}_0\right)} \quad \text{(A.33)}$$

With the identities $(\mathbf{A}\otimes\mathbf{B})^{-1} = \mathbf{A}^{-1}\otimes\mathbf{B}^{-1}$ and $(\mathbf{A}\otimes\mathbf{B})^T = \mathbf{A}^T\otimes\mathbf{B}^T$, the equation becomes

$$f(\underline{\mu}_s \mid \mathbf{C}_s,\boldsymbol{\Theta},\mathbf{X}_s) \propto e^{-\frac{1}{2}\left(\underline{1}_{m\times1}^T\otimes\underline{\mu}_s^T\right)\left(\mathbf{R}^{-1}\otimes\mathbf{C}_s^{-1}\right)\left(\underline{1}_{m\times1}\otimes\underline{\mu}_s\right)+\underline{\mathbf{X}}_s^T\left(\mathbf{R}^{-1}\otimes\mathbf{C}_s^{-1}\right)\left(\underline{1}_{m\times1}\otimes\underline{\mu}_s\right)-\frac{1}{2}\left(\underline{\mu}_s-\underline{\mu}_0\right)^T\mathbf{C}_0^{-1}\left(\underline{\mu}_s-\underline{\mu}_0\right)} \quad \text{(A.34)}$$

This is a multivariate normal PDF of $\underline{\mu}_s$ because $\ln[f(\underline{\mu}_s|\mathbf{C}_s,\boldsymbol{\Theta},\mathbf{X}_s)]$ is quadratic in $\underline{\mu}_s$. The mean vector of this multivariate normal PDF can be found by solving

$$\frac{\partial \ln\left[f\left(\underline{\mu}_s \mid C_s, \Theta, X_s\right)\right]}{\partial \underline{\mu}_s}$$

$$= \frac{\partial\left[-\frac{1}{2}\left(\underline{1}_{m\times1}^T \otimes \underline{\mu}_s^T\right)\left(\mathbf{R}^{-1} \otimes \mathbf{C}_s^{-1}\right)\left(\underline{1}_{m\times1} \otimes \underline{\mu}_s\right) + \underline{\mathbf{X}}_s^T\left(\mathbf{R}^{-1} \otimes \mathbf{C}_s^{-1}\right)\left(\underline{1}_{m\times1} \otimes \underline{\mu}_s\right) - \frac{1}{2}\left(\underline{\mu}_s - \underline{\mu}_0\right)^T \mathbf{C}_0^{-1}\left(\underline{\mu}_s - \underline{\mu}_0\right)\right]}{\partial \mu_s}$$

$$= -\left(\underline{1}_{m\times1}^T \otimes \underline{\mu}_s^T\right)\left(\mathbf{R}^{-1} \otimes \mathbf{C}_s^{-1}\right)\left(\underline{1}_{m\times1} \otimes \mathbf{I}_{n\times n}\right) + \underline{\mathbf{X}}_s^T\left(\mathbf{R}^{-1} \otimes \mathbf{C}_s^{-1}\right)\left(\underline{1}_{m\times1} \otimes \mathbf{I}_{n\times n}\right) - \left(\underline{\mu}_s - \underline{\mu}_0\right)^T \mathbf{C}_0^{-1} = \underline{0}_{1\times n} \quad \text{(A.35)}$$

where $\underline{0}_{1\times n}$ is a zero vector of size (1×n). Apply the identity $(\mathbf{A}\otimes\mathbf{B})(\mathbf{C}\otimes\mathbf{D}) = (\mathbf{AC})\otimes(\mathbf{BD})$ and take the matrix transpose, we have

$$-\left(\underline{1}_{m\times1}^T \mathbf{R}^{-1} \underline{1}_{m\times1}\right) \otimes \left(\mathbf{C}_s^{-1}\underline{\mu}_s\right) + \left[\left(\underline{1}_{m\times1}^T \mathbf{R}^{-1}\right) \otimes \mathbf{C}_s^{-1}\right]\underline{\mathbf{X}}_s - \mathbf{C}_0^{-1}\left(\underline{\mu}_s - \underline{\mu}_0\right) = \underline{0}_{n\times1} \quad \text{(A.36)}$$

Note that $\underline{1}^T_{m\times1}\mathbf{R}^{-1}\underline{1}_{m\times1}$ is a scalar and that $a\otimes\mathbf{B} = a\times\mathbf{B}$. Equation (A.36) becomes

$$\left[\mathbf{C}_0^{-1} + \left(\underline{1}_{m\times1}^T \mathbf{R}^{-1} \underline{1}_{m\times1}\right)\mathbf{C}_s^{-1}\right]\underline{\mu}_s = \mathbf{C}_s^{-1}\underline{\mu}_0 + \left[\left(\underline{1}_{m\times1}^T \mathbf{R}^{-1}\right) \otimes \mathbf{C}_s^{-1}\right]\underline{\mathbf{X}}_s \quad \text{(A.37)}$$

Therefore, the mean vector of the multivariate normal PDF $f(\underline{\mu}_s|\mathbf{C}_s,\boldsymbol{\Theta},\mathbf{X}_s)$ is

$$E(\underline{\mu}_s \mid \mathbf{C}_s, \Theta, \mathbf{X}_s) = \left[\mathbf{C}_0^{-1} + \left(\underline{1}_{m\times1}^T \mathbf{R}^{-1} \underline{1}_{m\times1}\right)\mathbf{C}_s^{-1}\right]^{-1}\left(\mathbf{C}_s^{-1}\underline{\mu}_0 + \left[\left(\underline{1}_{m\times1}^T \mathbf{R}^{-1}\right) \otimes \mathbf{C}_s^{-1}\right]\underline{\mathbf{X}}_s\right) \quad \text{(A.38)}$$

By replacing $\mathbf{R} = \mathbf{R}_h\otimes\mathbf{R}_z$ into the above equation, we have

$$\begin{aligned}
&E(\underline{\mu}_s \mid \mathbf{C}_s, \Theta, \mathbf{X}_s) \\
&= \left[\mathbf{C}_0^{-1} + \left(\underline{1}_{m\times1}^T \left(\mathbf{R}_h \otimes \mathbf{R}_z\right)^{-1} \underline{1}_{m\times1}\right)\mathbf{C}_s^{-1}\right]^{-1}\left(\mathbf{C}_s^{-1}\underline{\mu}_0 + \left[\left(\underline{1}_{m\times1}^T \left(\mathbf{R}_h \otimes \mathbf{R}_z\right)^{-1}\right) \otimes \mathbf{C}_s^{-1}\right]\underline{\mathbf{X}}_s\right) \\
&= \left[\mathbf{C}_0^{-1} + \left(\left(\underline{1}_{m_h\times1}^T \otimes \underline{1}_{m_z\times1}^T\right)\left(\mathbf{R}_h^{-1} \otimes \mathbf{R}_z^{-1}\right)\left(\underline{1}_{m_h\times1} \otimes \underline{1}_{m_z\times1}\right)\right)\mathbf{C}_s^{-1}\right]^{-1} \\
&\quad \left(\mathbf{C}_s^{-1}\underline{\mu}_0 + \left[\left(\mathrm{vec}\left(\mathbf{1}_{m_z\times m_h}\right)^T\left(\mathbf{R}_h^{-1} \otimes \mathbf{R}_z^{-1}\right)\right) \otimes \mathbf{C}_s^{-1}\right]\underline{\mathbf{X}}_s\right) \\
&= \left[\mathbf{C}_0^{-1} + \left(\underline{1}_{m_h\times1}^T \mathbf{R}_h^{-1} \underline{1}_{m_h\times1}\right)\left(\underline{1}_{m_z\times1}^T \mathbf{R}_z^{-1} \underline{1}_{m_z\times1}\right)\mathbf{C}_s^{-1}\right]^{-1} \\
&\quad \left(\mathbf{C}_s^{-1}\underline{\mu}_0 + \left[\mathrm{vec}\left(\mathbf{R}_z^{-1}\mathbf{1}_{m_z\times m_h}\mathbf{R}_h^{-1}\right)^T \otimes \mathbf{C}_s^{-1}\right]\underline{\mathbf{X}}_s\right) \qquad \text{(A.39)} \\
&= \left[\mathbf{C}_0^{-1} + \left(\underline{1}_{m_h\times1}^T \mathbf{R}_h^{-1} \underline{1}_{m_h\times1}\right)\left(\underline{1}_{m_z\times1}^T \mathbf{R}_z^{-1} \underline{1}_{m_z\times1}\right)\mathbf{C}_s^{-1}\right]^{-1} \\
&\quad \left[\mathbf{C}_s^{-1}\underline{\mu}_0 + \mathbf{C}_s^{-1} \cdot \mathrm{mat}\left(\underline{\mathbf{X}}_s\right) \cdot \mathrm{vec}\left(\mathbf{R}_z^{-1} \cdot \mathbf{1}_{m_z\times m_h} \cdot \mathbf{R}_h^{-1}\right)\right]
\end{aligned}$$

In the above equation, we have considered the identity $(\mathbf{A}^T \otimes \mathbf{B}) \times \text{vec}(\mathbf{C}) = \text{vec}(\mathbf{BCA})$. The covariance matrix of $f(\underline{\mu}_s | \mathbf{C}_s, \boldsymbol{\Theta}, \mathbf{X}_s)$ can be obtained by taking the inverse of the negative Hessian matrix for $\ln[f(\underline{\mu}_s | \mathbf{C}_s, \boldsymbol{\Theta}, \mathbf{X}_s)]$:

$$\left\{-\nabla^2 \ln\left[f\left(\underline{\mu}_s | \mathbf{C}_s, \Theta, \mathbf{X}_s\right)\right]\right\}^{-1} = \left(-\frac{\partial\left\{\partial \ln\left[f\left(\underline{\mu}_s | \mathbf{C}_s, \Theta, \mathbf{X}_s\right)\right] / \partial \underline{\mu}_s\right\}^T}{\partial \underline{\mu}_s}\right)^{-1}$$

$$= \left(-\frac{\partial\left\{\left[-\left(\underline{1}_{m\times1}^T \otimes \underline{\mu}_s^T\right)\left(\mathbf{R}^{-1} \otimes \mathbf{C}_s^{-1}\right)\left(\underline{1}_{m\times1} \otimes \mathbf{I}_{n\times n}\right) + \underline{\mathbf{X}}_s^T\left(\mathbf{R}^{-1} \otimes \mathbf{C}_s^{-1}\right)\left(\underline{1}_{m\times1} \otimes \mathbf{I}_{n\times n}\right) - \left(\underline{\mu}_s - \underline{\mu}_0\right)^T \mathbf{C}_0^{-1}\right]^T\right\}}{\partial \underline{\mu}_s}\right)^{-1}$$

$$= \left(\frac{\partial\left[\mathbf{C}_0^{-1}\left(\underline{\mu}_s - \underline{\mu}_0\right) + \left(\underline{1}_{m\times1}^T \mathbf{R}^{-1} \underline{1}_{m\times1}\right) \mathbf{C}_s^{-1} \underline{\mu}_s\right]}{\partial \underline{\mu}_s}\right)^{-1} = \left[\mathbf{C}_0^{-1} + \left(\underline{1}_{m\times1}^T \mathbf{R}^{-1} \underline{1}_{m\times1}\right) \mathbf{C}_s^{-1}\right]^{-1}$$

$$= \left[\mathbf{C}_0^{-1} + \left(\underline{1}_{m_h\times1}^T \mathbf{R}_h^{-1} \underline{1}_{m_h\times1}\right)\left(\underline{1}_{m_z\times1}^T \mathbf{R}_z^{-1} \underline{1}_{m_z\times1}\right) \mathbf{C}_s^{-1}\right]^{-1} \tag{A.40}$$

As a result, $f(\underline{\mu}_s | \mathbf{C}_s, \boldsymbol{\Theta}, \mathbf{X}_s)$ is a multivariate normal PDF with the following mean vector and covariance matrix:

$$\begin{aligned} &E(\underline{\mu}_s | \mathbf{C}_s, \Theta, \mathbf{X}_s) \\ &\quad = \text{Var}(\underline{\mu}_s | \mathbf{C}_s, \Theta, \mathbf{X}_s) \cdot \left[\mathbf{C}_s^{-1} \underline{\mu}_0 + \mathbf{C}_s^{-1} \cdot \text{mat}\left(\underline{\mathbf{X}}_s\right) \cdot \text{vec}\left(\mathbf{R}_z^{-1} \cdot \mathbf{1}_{m_z \times m_h} \cdot \mathbf{R}_h^{-1}\right)\right] \\ &\text{Var}(\underline{\mu}_s | \mathbf{C}_s, \Theta, \mathbf{X}_s) = \left[\mathbf{C}_0^{-1} + \left(\underline{1}_{m_h\times1}^T \mathbf{R}_h^{-1} \underline{1}_{m_h\times1}\right)\left(\underline{1}_{m_z\times1}^T \mathbf{R}_z^{-1} \underline{1}_{m_z\times1}\right) \mathbf{C}_s^{-1}\right]^{-1} \end{aligned} \tag{A.41}$$

For the full conditional PDF $f(\mathbf{C}_s | \underline{\mu}_s, \boldsymbol{\Theta}, \mathbf{X}_s)$, one has to focus on the $\mathbf{C}_s$-related terms in Equation (A.32):

$$f(\mathbf{C}_s | \underline{\mu}_s, \Theta, \mathbf{X}_s) \propto \left|\mathbf{R} \otimes \mathbf{C}_s\right|^{-\frac{1}{2}} \cdot \left|\mathbf{C}_s\right|^{-\frac{v_0+n+1}{2}} e^{-\frac{1}{2}\left(\underline{\mathbf{X}}_s - \underline{1}_{m\times1} \otimes \underline{\mu}_s\right)^T (\mathbf{R} \otimes \mathbf{C}_s)^{-1}\left(\underline{\mathbf{X}}_s - \underline{1}_{m\times1} \otimes \underline{\mu}_s\right) - \frac{1}{2}\text{tr}\left(\Sigma_0 \times \mathbf{C}_s^{-1}\right)} \tag{A.42}$$

Consider the Cholesky decompositions for **R** and $\mathbf{C}_s$: $\mathbf{R} = \mathbf{L}_R \times \mathbf{L}_R^T$ and $\mathbf{C}_s = \mathbf{L}_C \times \mathbf{L}_C^T$. Also consider the identities $|\mathbf{A}_{m\times m} \otimes \mathbf{B}_{n\times n}| = |\mathbf{A}|^n \times |\mathbf{B}|^m$ and $\underline{\mathbf{X}}_s - \underline{1} \otimes \underline{\mu}_s = \text{vec}[\text{mat}(\underline{\mathbf{X}}_s) - \underline{\mu}_s \times \underline{1}^T]$, where $\text{mat}(\underline{\mathbf{X}}_s)$ is an $(n\times m)$ matrix formed by stacking the column vectors $\underline{\mathbf{X}}_s^{(1)}$, $\underline{\mathbf{X}}_s^{(2)}$, …, and $\underline{\mathbf{X}}_s^{(m)}$ horizontally, and vec(.) is the inverse process of mat(.): $\text{vec}(\mathbf{A}_{n\times m})$ stacks the column vectors of matrix **A** into an $(nm\times1)$ column vector. Equation (A.42) becomes

$$\begin{aligned} f(\mathbf{C}_s | \underline{\mu}_s, \Theta, \mathbf{X}_s) &\propto \left|\mathbf{C}_s\right|^{-\frac{m+v_0+n+1}{2}} e^{-\frac{1}{2}\text{vec}\left[\text{mat}(\underline{\mathbf{X}}_s)\ \underline{\mu}_s \underline{1}_{m\times1}^T\right]^T\left[\left(\mathbf{L}_R \mathbf{L}_R^T\right)^{-1} \otimes \left(\mathbf{L}_C \mathbf{L}_C^T\right)^{-1}\right]\text{vec}\left[\text{mat}(\underline{\mathbf{X}}_s)\ \underline{\mu}_s \underline{1}_{m\times1}^T\right] - \frac{1}{2}\text{tr}\left(\Sigma_0 \times \mathbf{C}_s^{-1}\right)} \\ &\propto \left|\mathbf{C}_s\right|^{-\frac{m+v_0+n+1}{2}} e^{-\frac{1}{2}\text{vec}\left[\text{mat}(\underline{\mathbf{X}}_s) - \underline{\mu}_s \underline{1}_{m\times1}^T\right]^T\left(\mathbf{L}_R^{-1} \otimes \mathbf{L}_C^{-1}\right)^T\left(\mathbf{L}_R^{-1} \otimes \mathbf{L}_C^{-1}\right)\text{vec}\left[\text{mat}(\underline{\mathbf{X}}_s) - \underline{\mu}_s \underline{1}_{m\times1}^T\right] - \frac{1}{2}\text{tr}\left(\Sigma_0 \times \mathbf{C}_s^{-1}\right)} \end{aligned} \tag{A.43}$$

Consider the identity $(\mathbf{A}^T \otimes \mathbf{B}) \times \text{vec}(\mathbf{C}) = \text{vec}(\mathbf{BCA})$. Therefore, $(\mathbf{LR}^{-1} \otimes \mathbf{LC}^{-1}) \times \text{vec}[\text{mat}(\underline{X}_s) - \underline{\mu}_s \times \underline{1}^T] = \text{vec}\{\mathbf{LC}^{-1} \times [\text{mat}(\underline{X}_s) - \underline{\mu}_s \times \underline{1}^T] \times \mathbf{LR}^{-T}\}$, where $\mathbf{A}^{-T}$ denotes the transpose of $\mathbf{A}^{-1}$. As a result, we have

$$f(C_s \mid \underline{\mu}_s, \boldsymbol{\Theta}, \mathbf{X}_s) \propto \left|C_s\right|^{-\frac{m+v_0+n+1}{2}} e^{-\frac{1}{2}\text{vec}\left(L_C^{-1}\left[\text{mat}(\underline{X}_s)-\underline{\mu}_s \underline{1}_{m\times1}^T\right]L_R^{-T}\right)^T \text{vec}\left(L_C^{-1}\left[\text{mat}(\underline{X}_s)-\underline{\mu}_s \underline{1}_{m\times1}^T\right]L_R^{-T}\right)-\frac{1}{2}\text{tr}\left(\Sigma_0 \times C_s^{-1}\right)} \tag{A.44}$$

Considering the identity $\text{vec}(\mathbf{A})^T\text{vec}(\mathbf{B}) = \text{tr}(\mathbf{A}^T\mathbf{B})$, we have

$$\begin{aligned} f(C_s \mid \underline{\mu}_s, \boldsymbol{\Theta}, \mathbf{X}_s) &\propto \left|C_s\right|^{-\frac{m+v_0+n+1}{2}} e^{-\frac{1}{2}\text{tr}\left[\left(L_C^{-1}\left[\text{mat}(\underline{X}_s)-\underline{\mu}_s \underline{1}_{m\times1}^T\right]L_R^{-T}\right)^T\left(L_C^{-1}\left[\text{mat}(\underline{X}_s)-\underline{\mu}_s \underline{1}_{m\times1}^T\right]L_R^{-T}\right)\right]-\frac{1}{2}\text{tr}\left(\Sigma_0 \times C_s^{-1}\right)} \\ &\propto \left|C_s\right|^{-\frac{m+v_0+n+1}{2}} e^{-\frac{1}{2}\text{tr}\left(L_R^{-1}\left[\text{mat}(\underline{X}_s)-\underline{\mu}_s \underline{1}_{m\times1}^T\right]^T L_C^{-T} \times L_C^{-1}\left[\text{mat}(\underline{X}_s)-\underline{\mu}_s \underline{1}_{m\times1}^T\right]L_R^{-T}\right)-\frac{1}{2}\text{tr}\left(\Sigma_0 \times C_s^{-1}\right)} \\ &\propto \left|C_s\right|^{-\frac{m+v_0+n+1}{2}} e^{-\frac{1}{2}\text{tr}\left(L_R^{-1}\left[\text{mat}(\underline{X}_s)-\underline{\mu}_s \underline{1}_{m\times1}^T\right]^T C_s^{-1}\left[\text{mat}(\underline{X}_s)-\underline{\mu}_s \underline{1}_{m\times1}^T\right]L_R^{-T}\right)-\frac{1}{2}\text{tr}\left(\Sigma_0 \times C_s^{-1}\right)} \end{aligned} \tag{A.45}$$

Considering the identity $\text{tr}(\mathbf{AB}) = \text{tr}(\mathbf{BA})$, we have

$$\begin{aligned} f(C_s \mid \underline{\mu}_s, \boldsymbol{\Theta}, \mathbf{X}_s) &\propto \left|C_s\right|^{-\frac{m+v_0+n+1}{2}} e^{-\frac{1}{2}\text{tr}\left(\left[\text{mat}(\underline{X}_s)-\underline{\mu}_s \underline{1}_{m\times1}^T\right]L_R^{-T}L_R^{-1}\left[\text{mat}(\underline{X}_s)-\underline{\mu}_s \underline{1}_{m\times1}^T\right]^T C_s^{-1}\right)-\frac{1}{2}\text{tr}\left(\Sigma_0 \times C_s^{-1}\right)} \\ &\propto \left|C_s\right|^{-\frac{m+v_0+n+1}{2}} e^{-\frac{1}{2}\text{tr}\left(\left[\text{mat}(\underline{X}_s)-\underline{\mu}_s \underline{1}_{m\times1}^T\right]R^{-1}\left[\text{mat}(\underline{X}_s)-\underline{\mu}_s \underline{1}_{m\times1}^T\right]^T C_s^{-1}\right)-\frac{1}{2}\text{tr}\left(\Sigma_0 \times C_s^{-1}\right)} \\ &\propto \left|C_s\right|^{-\frac{(m+v_0)+n+1}{2}} e^{-\frac{1}{2}\text{tr}\left(\left[\Sigma_0+\left[\text{mat}(\underline{X}_s)-\underline{\mu}_s \underline{1}_{m\times1}^T\right]R^{-1}\left[\text{mat}(\underline{X}_s)-\underline{\mu}_s \underline{1}_{m\times1}^T\right]^T\right]C_s^{-1}\right)} \end{aligned} \tag{A.46}$$

This is an inverse-Wishart PDF with scale matrix $= \boldsymbol{\Sigma}_0 + [\text{mat}(\underline{X}_s) - \underline{\mu}_s \times \underline{1}^T]\mathbf{R}^{-1}[\text{mat}(\underline{X}_s) - \underline{\mu}_s \times \underline{1}^T]^T$ and degree of freedom $= m+v_0$:

$$f(C_s \mid \underline{\mu}_s, \boldsymbol{\Theta}, \mathbf{X}_s) = \text{IW}\left\{\Sigma_0 + \left[\text{mat}(\underline{X}_s) - \underline{\mu}_s \underline{1}_{m\times1}^T\right]\mathbf{R}^{-1}\left[\text{mat}(\underline{X}_s) - \underline{\mu}_s \underline{1}_{m\times1}^T\right]^T, m + v_0\right\} \tag{A.47}$$

By replacing $\mathbf{R} = \mathbf{R}_h \otimes \mathbf{R}_z$ into the above equation, we have

$$\begin{aligned} &f(C_s \mid \underline{\mu}_s, \boldsymbol{\Theta}, \mathbf{X}_s) \\ &= \text{IW}\left\{\Sigma_0 + \left[\text{mat}(\underline{X}_s) - \underline{\mu}_s \underline{1}_{m\times1}^T\right]\left(\mathbf{R}_h^{-1} \otimes \mathbf{R}_z^{-1}\right)\left[\text{mat}(\underline{X}_s) - \underline{\mu}_s \underline{1}_{m\times1}^T\right]^T, m + v_0\right\} \\ &= \text{IW}\left\{\Sigma_0 + \mathbf{S}, m + v_0\right\} \end{aligned} \tag{A.48}$$

where $\mathbf{S} \in \mathbf{R}^{n\times n}$ and its (i,j) entry S_{ij} can be computed as

$$S_{ij} = \text{trace}\left[\text{mat}\left(\underline{X}_{s,i.} - \mu_{s,i}\underline{1}_{m\times1}^T\right)^T \cdot \mathbf{R}_z^{-1} \cdot \text{mat}\left(\underline{X}_{s,j.} - \mu_{s,j}\underline{1}_{m\times1}^T\right) \cdot \mathbf{R}_h^{-1}\right] \tag{A.49}$$

where $\underline{X}_{s,i,\cdot}\in R^{1\times m}$ is the ith row of mat($\underline{X}_s$), containing the ith soil property for all m locations; $\mu_{s,i}$ is the ith entry of $\underline{\mu}_s$, which is the mean value for the ith soil property.

APPENDIX 5.3: DERIVATIONS FOR $f(X_{hz'}|\underline{\mu}_s, C_s, X_{hz})$ AND $f(X_{h'(Z\cup Z')}|\underline{\mu}_s, C_s, X_{h(Z\cup Z')})$

The derivations for $f(X_{hz'}|\underline{\mu}_s, C_s, X_{hz})$ are as follows. First, note that given ($\underline{\mu}_s$, C_s), vec($X_{hz'}$) and vec(X_{hz}) are jointly normal with the following mean vectors and covariance matrices:

$$E\left(\begin{bmatrix}\mathrm{vec}(X_{hz'})\\ \mathrm{vec}(X_{hz})\end{bmatrix} \mid \underline{\mu}_s, C_s\right) = \begin{bmatrix}\underline{1}_{(m_h\times m_{z'})\times 1}\otimes\underline{\mu}_s\\ \underline{1}_{(m_h\times m_z)\times 1}\otimes\underline{\mu}_s\end{bmatrix}$$
$$COV\left(\begin{bmatrix}\mathrm{vec}(X_{hz'})\\ \mathrm{vec}(X_{hz})\end{bmatrix} \mid \underline{\mu}_s, C_s\right) = \begin{bmatrix}R_h\otimes R_{z'}\otimes C_s & R_h\otimes R_{z'z}\otimes C_s\\ R_h\otimes R_{zz'}\otimes C_s & R_h\otimes R_z\otimes C_s\end{bmatrix} \quad (A.50)$$

where E(.) denotes a mean vector; COV(.,.) denotes a covariance matrix. It can be shown that conditioning on X_{hz}, vec($X_{hz'}$) is still multivariate normal with the following mean vector and covariance matrix:

$$E(\mathrm{vec}(X_{hz'}) \mid \underline{\mu}_s, C_s, X_{hz})$$
$$= \underline{1}_{(m_h\times m_{z'})\times 1}\otimes\underline{\mu}_s + (R_h\otimes R_{z'z}\otimes C_s)(R_h\otimes R_z\otimes C_s)^{-1}\left[\mathrm{vec}(X_{hz}) - \underline{1}_{(m_h\times m_z)\times 1}\otimes\underline{\mu}_s\right]$$
$$= \underline{1}_{(m_h\times m_{z'})\times 1}\otimes\underline{\mu}_s + \left(I_{m_h\times m_h}\otimes R_{z'z}R_z^{-1}\otimes I_{n\times n}\right)\left[\mathrm{vec}(X_{hz}) - \underline{1}_{(m_h\times m_z)\times 1}\otimes\underline{\mu}_s\right]$$
$$= \mathrm{vec}\left[\underline{\mu}_s\times\underline{1}^T_{(m_h\times m_{z'})\times 1} + \left(X_{hz} - \underline{\mu}_s\times\underline{1}^T_{(m_h\times m_z)\times 1}\right)\left(I_{m_h\times m_h}\otimes R_z^{-T}R_{z'z}^T\right)\right]$$
$$COV(\mathrm{vec}(X_{hz'}) \mid \underline{\mu}_s, C_s, X_{hz}) = (R_h\otimes R_{z'}\otimes C_s)$$
$$-(R_h\otimes R_{z'z}\otimes C_s)(R_h\otimes R_z\otimes C_s)^{-1}(R_h\otimes R_{zz'}\otimes C_s)$$
$$= (R_h\otimes R_{z'}\otimes C_s) - \left[R_h\otimes\left(R_{z'z}R_z^{-1}R_{zz'}\right)\otimes C_s\right] = R_h\otimes\left(R_{z'} - R_{z'z}R_z^{-1}R_{zz'}\right)\otimes C_s \quad (A.51)$$

If we further define $R_{z'|z} \equiv R_{z'} - R_{z'z}\times R_z^{-1}\times R_{zz'}$, it is clear that

$$E(\mathrm{vec}(X_{hz'}) \mid \underline{\mu}_s, C_s, X_{hz})$$
$$= \mathrm{vec}\left[\underline{\mu}_s\times\underline{1}^T_{(m_h\times m_{z'})\times 1} + \left(X_{hz} - \underline{\mu}_s\times\underline{1}^T_{(m_h\times m_z)\times 1}\right)\left(I_{m_h\times m_h}\otimes R_z^{-T}R_{z'z}^T\right)\right] \quad (A.52)$$
$$COV(\mathrm{vec}(X_{hz'}) \mid \underline{\mu}_s, C_s, X_{hz}) = R_h\otimes R_{z'|z}\otimes C_s$$

It is then clear that a sample of $\text{vec}(\mathbf{X}_{hz'})$ can be drawn by the following equation:

$$
\begin{aligned}
\text{vec}\left(\mathbf{X}_{hz'}\right) &= \text{vec}\left[\underline{\mu}_s \times \underline{1}^T_{(m_h \times m_{z'})\times 1} + \left(\mathbf{X}_{hz} - \underline{\mu}_s \times \underline{1}^T_{(m_h \times m_z)\times 1}\right)\left(\mathbf{I}_{m_h \times m_h} \otimes \mathbf{R}_z^{-T}\mathbf{R}_{z'z}^T\right)\right] \\
&\quad + \left(\mathbf{L}_h \otimes \mathbf{L}_{z'|z} \otimes \mathbf{L}_s\right)\text{vec}\left(\mathbf{Z}_{n\times(m_h \times m_{z'})}\right) \\
&= \text{vec}\left[\underline{\mu}_s \times \underline{1}^T_{(m_h \times m_{z'})\times 1} + \left(\mathbf{X}_{hz} - \underline{\mu}_s \times \underline{1}^T_{(m_h \times m_z)\times 1}\right)\left(\mathbf{I}_{m_h \times m_h} \otimes \mathbf{R}_z^{-T}\mathbf{R}_{z'z}^T\right)\right] \\
&\quad + \text{vec}\left(\mathbf{L}_s\mathbf{Z}_{n\times(m_h \times m_{z'})}\left(\mathbf{L}_h^T \otimes \mathbf{L}_{z'|z}^T\right)\right) \\
\mathbf{X}_{hz'} &= \underline{\mu}_s \times \underline{1}^T_{(m_h \times m_{z'})\times 1} + \left(\mathbf{X}_{hz} - \underline{\mu}_s \times \underline{1}^T_{(m_h \times m_z)\times 1}\right)\left(\mathbf{I}_{m_h \times m_h} \otimes \mathbf{R}_z^{-T}\mathbf{R}_{z'z}^T\right) \\
&\quad + \mathbf{L}_s\mathbf{Z}_{n\times(m_h \times m_{z'})}\left(\mathbf{L}_h^T \otimes \mathbf{L}_{z'|z}^T\right)
\end{aligned}
\tag{A.53}
$$

where $\mathbf{L}_h$, $\mathbf{L}_{z'|z}$, and $\mathbf{L}_s$ are the lower Cholesky decompositions of $\mathbf{R}_h$, $\mathbf{R}_{z'|z}$, and $\mathbf{C}_s$, respectively. By removing the vec(.) sign, we have

$$
\begin{aligned}
\mathbf{X}_{hz'} &= \underline{\mu}_s \times \underline{1}^T_{(m_h \times m_{z'})\times 1} + (\mathbf{X}_{hz} - \underline{\mu}_s \times \underline{1}^T_{(m_h \times m_z)\times 1})\left(\mathbf{I}_{m_h \times m_h} \otimes \mathbf{R}_z^{-T}\mathbf{R}_{z'z}^T\right) \\
&\quad + \mathbf{L}_s\mathbf{Z}_{n\times(m_h \times m_{z'})}\left(\mathbf{L}_h^T \otimes \mathbf{L}_{z'|z}^T\right)
\end{aligned}
\tag{A.54}
$$

which is the same as Equation (5.27).

The derivations for $f(\mathbf{X}_{hz'}|\underline{\mu}_s, \mathbf{C}_s, \mathbf{X}_{h(z\cup z')})$ are similar to the above. Conditioning on $\mathbf{X}_{h(z\cup z')}$, $\text{vec}(\mathbf{X}_{h'(z\cup z')})$ is still multivariate normal with the following mean vector and covariance matrix:

$$
\begin{aligned}
&E(\text{vec}\left(\mathbf{X}_{h'(z\cup z')}\right) \mid \underline{\mu}_s, \mathbf{C}_s, \mathbf{X}_{h(z\cup z')}) \\
&= \underline{1}_{[m_{h'}\times(m_z+m_{z'})]\times 1} \otimes \underline{\mu}_s + (\mathbf{R}_{h'h} \otimes \mathbf{R}_{z\cup z'} \otimes \mathbf{C}_s)(\mathbf{R}_h \otimes \mathbf{R}_{z\cup z'} \otimes \mathbf{C}_s)^{-1} \\
&\quad \left[\text{vec}\left(\mathbf{X}_{h(z\cup z')}\right) - \underline{1}_{[m_h\times(m_z+m_{z'})]\times 1} \otimes \underline{\mu}_s\right] \\
&= \underline{1}_{[m_{h'}\times(m_z+m_{z'})]\times 1} \otimes \underline{\mu}_s + \left(\mathbf{R}_{h'h}\mathbf{R}_h^{-1} \otimes \mathbf{I}_{(m_z+m_{z'})\times(m_z+m_{z'})} \otimes \mathbf{I}_n\right) \\
&\quad \left[\text{vec}\left(\mathbf{X}_{h(z\cup z')}\right) - \underline{1}_{[m_h\times(m_z+m_{z'})]\times 1} \otimes \underline{\mu}_s\right] \\
&= \text{vec}\left[\underline{\mu}_s \times \underline{1}^T_{[m_{h'}\times(m_z+m_{z'})]\times 1} + \left(\mathbf{X}_{h(z\cup z')} - \underline{\mu}_s \times \underline{1}^T_{[m_h\times(m_z+m_{z'})]\times 1}\right)\left(\mathbf{R}_h^{-T}\mathbf{R}_{h'h}^T \otimes \mathbf{I}_{(m_z+m_{z'})\times(m_z+m_{z'})}\right)\right] \\
&\quad \text{COV}(\text{vec}\left(\mathbf{X}_{h'(z\cup z')}\right) \mid \underline{\mu}_s, \mathbf{C}_s, \mathbf{X}_{h(z\cup z')}) = \left(\mathbf{R}_{h'} \otimes \mathbf{R}_{z\cup z'} \otimes \mathbf{C}_s\right) \\
&\quad - \left(\mathbf{R}_{h'h} \otimes \mathbf{R}_{z\cup z'} \otimes \mathbf{C}_s\right)\left(\mathbf{R}_h \otimes \mathbf{R}_{z\cup z'} \otimes \mathbf{C}_s\right)^{-1}\left(\mathbf{R}_{hh'} \otimes \mathbf{R}_{z\cup z'} \otimes \mathbf{C}_s\right) \\
&= \left(\mathbf{R}_{h'} \otimes \mathbf{R}_{z\cup z'} \otimes \mathbf{C}_s\right) - \left(\left(\mathbf{R}_{h'h}\mathbf{R}_h^{-1}\mathbf{R}_{hh'}\right) \otimes \mathbf{R}_{z\cup z'} \otimes \mathbf{C}_s\right) = \left(\mathbf{R}_{h'} - \mathbf{R}_{h'h}\mathbf{R}_h^{-1}\mathbf{R}_{hh'}\right) \otimes \mathbf{R}_{z\cup z'} \otimes \mathbf{C}_s
\end{aligned}
\tag{A.55}
$$

If we further define $\mathbf{R}_{h'|h} \equiv \mathbf{R}_{h'} - \mathbf{R}_{h'h} \times \mathbf{R}_h^{-1} \times \mathbf{R}_{hh'}$, it is clear that

$$E\left(\text{vec}\left(\mathbf{X}_{h'(z\cup z')}\right) \mid \underline{\mu}_s, \mathbf{C}_s, \mathbf{X}_{h(z\cup z')}\right)$$
$$= \text{vec}\left[\underline{\mu}_s \times \underline{1}^T_{[m_{h'}\times(m_z+m_{z'})]\times 1} + \left(\mathbf{X}_{h(z\cup z')} - \underline{\mu}_s \times \underline{1}^T_{[m_h\times(m_z+m_{z'})]\times 1}\right)\left(\mathbf{R}_h^{-T}\mathbf{R}_{h'h}^T \otimes \mathbf{I}_{(m_z+m_{z'})\times(m_z+m_{z'})}\right)\right]$$
$$\text{COV}\left(\text{vec}\left(\mathbf{X}_{h'(z\cup z')}\right) \mid \underline{\mu}_s, \mathbf{C}_s, \mathbf{X}_{h(z\cup z')}\right) = \mathbf{R}_{h'|h} \otimes \mathbf{R}_{z\cup z'} \otimes \mathbf{C}_s \quad \text{(A.56)}$$

A sample of $\text{vec}(\mathbf{X}_{h(z\cup z')})$ can be drawn by the following equation:

$$\text{vec}\left(\mathbf{X}_{h'(z\cup z')}\right) = \text{vec}\left[\underline{\mu}_s \times \underline{1}^T_{[m_{h'}\times(m_z+m_{z'})]\times 1} + \left(\mathbf{X}_{h(z\cup z')} - \underline{\mu}_s \times \underline{1}^T_{[m_h\times(m_z+m_{z'})]\times 1}\right)\right.$$
$$\left.\left(\mathbf{R}_h^{-T}\mathbf{R}_{h'h}^T \otimes \mathbf{I}_{(m_z+m_{z'})\times(m_z+m_{z'})}\right)\right] + \left(\mathbf{L}_{h'|h} \otimes \mathbf{L}_{z\cup z'} \otimes \mathbf{L}_s\right)\text{vec}\left(\mathbf{Z}_{n\times[m_{h'}\times(m_z+m_{z'})]}\right)$$
$$= \text{vec}\left[\underline{\mu}_s \times \underline{1}^T_{[m_{h'}\times(m_z+m_{z'})]\times 1} + \left(\mathbf{X}_{h(z\cup z')} - \underline{\mu}_s \times \underline{1}^T_{[m_h\times(m_z+m_{z'})]\times 1}\right)\right.$$
$$\left.\left(\mathbf{R}_h^{-T}\mathbf{R}_{h'h}^T \otimes \mathbf{I}_{(m_z+m_{z'})\times(m_z+m_{z'})}\right) + \mathbf{L}_s\mathbf{Z}_{n\times[m_{h'}\times(m_z+m_{z'})]}\left(\mathbf{L}_{h'|h}^T \otimes \mathbf{L}_{z\cup z'}^T\right)\right] \quad \text{(A.57)}$$

where $\mathbf{L}_{h'|h}$ and $\mathbf{L}_{z'\cup z}$ are the lower Cholesky decompositions of $\mathbf{R}_{h'|h}$ and $\mathbf{R}_{z'\cup z}$, respectively. By removing the vec(.) sign, we have

$$\mathbf{X}_{h'(z\cup z')} = \underline{\mu}_s \times \underline{1}^T_{[m_{h'}\times(m_z+m_{z'})]\times 1} + \left(\mathbf{X}_{h(z\cup z')} - \underline{\mu}_s \times \underline{1}^T_{[m_h\times(m_z+m_{z'})]\times 1}\right)$$
$$\left(\mathbf{R}_h^{-T}\mathbf{R}_{h'h}^T \otimes \mathbf{I}_{(m_z+m_{z'})\times(m_z+m_{z'})}\right) + \mathbf{L}_s\mathbf{Z}_{n\times[m_{h'}\times(m_z+m_{z'})]}\left(\mathbf{L}_{h'|h}^T \otimes \mathbf{L}_{z\cup z'}^T\right) \quad \text{(A.58)}$$

which is the same as Equation (5.28).

References

Ang, A.H.-S. and W.H. Tang. 1975. *Probability concepts in engineering planning and design, Vol. 1, basic principles*. John Wiley, New York.

Ang, A.H.-S. and W.H. Tang. 2007. *Probability concepts in engineering*. John Wiley, New York.

Baecher, G.B. 2019. Putting numbers on geotechnical judgment. Companion white-paper to the 27th Buchanan Lecture, presented at Texas A&M University, College Station, October 18, 2019.

Baecher, G.B. and J.T. Christian. 2003. *Reliability and statistics in geotechnical engineering*. Wiley, Chichester.

Beck, J.L. 2010. Bayesian system identification based on probability logic. *Structural Control and Health Monitoring* 17: 825–847.

Beck, J.L. and S.K. Au. 2002. Bayesian updating of structural models and reliability using Markov chain Monte Carlo simulation. *Journal of Engineering Mechanics* 128(4): 380–391.

Beck, J.L. and L.S. Katafygiotis. 1998. Updating models and their uncertainties I: Bayesian statistical framework. *Journal of Engineering Mechanics* 124(4): 455–461.

Beck, J.L. and K.V. Yuen. 2004. Model selection using response measurements: Bayesian probabilistic approach. *Journal of Engineering Mechanics* 130(2): 192–203.

Benjamin, J.R. and C.A. Cornell. 1970. *Probability, statistics and decision for civil engineers*. McGraw-Hill, New York.

Cami, B., S. Javankhoshdel, K.K. Phoon, and J. Ching. 2020. Scale of fluctuation for spatially varying soils: Estimation methods and values. *ASCE-ASME Journal of Risk and Uncertainty in Engineering Systems, Part A: Civil Engineering* 6(4): 03120002.

Ching, J. and Y.C. Chen. 2007. Transitional Markov chain Monte Carlo method for Bayesian model updating, model class selection, and model averaging. *Journal of Engineering Mechanics* 133(7): 816–832.

Ching, J. and K.K. Phoon. 2014. Transformations and correlations among some clay parameters – the global database. *Canadian Geotechnical Journal* 51(6): 663–685.

Ching, J. and K.K. Phoon. 2020. Constructing a site-specific multivariate probability distribution using sparse, incomplete, and spatially variable (MUSIC-X) data. *Journal of Engineering Mechanics* 146(7): 04020061.

Ching, J., W.H. Huang, and K.K. Phoon. 2020. 3D probabilistic site characterization by sparse Bayesian learning. *Journal of Engineering Mechanics* 146(12): 04020134.

Ching, J., S. Wu, and K.K. Phoon. 2021a. Constructing quasi-site-specific multivariate probability distribution using hierarchical Bayesian model. *Journal of Engineering Mechanics* 147(10): 04021069.

Ching, J., Z.Y. Yang, and K.K. Phoon. 2021b. Dealing with non-lattice data in three-dimensional probabilistic site characterization. *Journal of Engineering Mechanics* 147(5): 06021003.

Ching, J., K.K. Phoon, Z.Y. Yang, and A.W. Stuedlein. 2022. Quasi-site-specific multivariate probability distribution model for sparse, incomplete, and three-dimensional spatially varying soil data. *Georisk: Assessment and Management of Risk for Engineered Systems and Geohazards* 16(1): 53–76.

Ching, J., I. Yoshida, and K.K. Phoon. 2023. Comparison of trend models for geotechnical spatial variability: Sparse Bayesian Learning vs. Gaussian Process Regression. *Gondwana Research* 123: 174–183.

Chung, Y., A. Gelman, S. Rabe-Hesketh, J. Liu, and V. Dorie. 2015. Weakly informative prior for point estimation of covariance matrices in hierarchical models. *Journal of Educational and Behavioral Statistics* 40(2): 136–157.

Conover, W.J. 1999. *Practical nonparametric statistics*, 3rd edition. John Wiley & Sons, New York.

Cox, R.T. 1946. Probability, frequency and reasonable expectation. *American Journal of Physics* 14: 1–13.

Cox, R.T. 1961. *The algebra of probable inference*. Johns Hopkins Press, Baltimore, MD.

Fenton, G.A. 1999. Random field modeling of CPT data. *Journal of Geotechnical and Geoenvironmental Engineering* 125(6): 486–498.

Fenton, G.A. and D.V. Griffiths. 2008. *Risk assessment in geotechnical engineering*. Wiley.

Fernandes, P., B. Plateau, and W.J. Stewart. 1998. Efficient descriptor-vector multiplications in stochastic automata networks. *Journal of the ACM* 45(3): 381–414.

Fink, D. 1997. A compendium of conjugate priors. http://www.people.cornell.edu/pages/df36/CONJINTRnew%20TEX.pdf

Gelman, A. and J. Hill. 2006. *Data analysis using regression and multilevel/hierarchical models*. Cambridge University Press.

Gelman, A., J.B. Carlin, H.S. Stern, D.B. Dunson, A. Vehtari, and D.B. Rubin. 2013. *Bayesian data analysis*, 3rd edition. Chapman and Hall/CRC.

Geman, S. and D. Geman. 1984. Stochastic relaxation, Gibbs distributions, and the Bayesian restoration of images. *IEEE Transactions on Pattern Analysis and Machine Intelligence* 6(6): 721–741.

Ghanem, R. and P.D. Spanos. 1991. *Stochastic finite element: A spectral approach*. Springer-Verlag, Berlin.

Gilks, W.R., S. Richardson, and D.J. Spiegelhalter. 1998. *Markov chain Monte Carlo in practice*. Chapman & Hall, Boca Raton, FL.

Guttorp, P. and T. Gneiting. 2006. Studies in the history of probability and statistics XLIX on the Matérn correlation family. *Biometrika* 93(4): 989–995.

Hastings, W.K. 1970. Monte Carlo sampling methods using Markov chains and their applications. *Biometrika* 57: 97–109.

Huang, A. and M.P. Wand. 2013. Simple marginally noninformative prior distributions for covariance matrices. *Bayesian Analysis* 8(2): 439–452.

ISSMGE-TC304. 2021. State-of-the-art review of inherent variability and uncertainty in geotechnical properties and models. International Society of Soil Mechanics and Geotechnical Engineering (ISSMGE) - Technical Committee TC304 'Engineering Practice of Risk Assessment and Management', March 2, 2021. http://140.112.12.21/issmge/2021/SOA_Review_on_geotechnical_property_variablity_and_model_uncertainty.pdf

Jaksa, M.B. 1995. The influence of spatial variability on the geotechnical design properties of a stiff, overconsolidated clay. Ph.D. thesis, University of Adelaide, Australia.

Jaksa, M.B., W.S. Kaggwa, and P.I. Brooker. 1999. Experimental evaluation of the scale of fluctuation of a stiff clay. *Proceedings of the 8th International Conference on Application of Statistics and Probability*, AA Balkema, Rotterdam, 415–422.

Jaynes, E.T. 1976. Confidence intervals vs Bayesian intervals. In *Foundations of Probability Theory, Statistical Inference, and Statistical Theories of Science*, Eds., W. L. Harper and C. A. Hooker, Springer, Dordrecht.

Johnson, N.L. 1949. Systems of frequency curves generated by methods of translation. *Biometrika* 36: 149–176.

Kulhawy, F.H. and P.W. Mayne. 1990. Manual on estimating soil properties for foundation design, Report EL-6800, Electric Power Research Institute, Cornell University, Palo Alto.

Liu, P.L. and A. Der Kiureghian. 1986. Multivariate distribution models with prescribed marginals and covariances. *Probabilistic Engineering Mechanics* 1(2): 105–112.

Lumb, P. 1966. The variability of natural soils. *Canadian Geotechnical Journal* 3(2): 74–97.

Mayne, P.W., B.R. Christopher, and J. DeJong. 2001. Manual on subsurface investigations. National Highway Institute Publication No. FHWA NHI-01-031, Federal Highway Administration, Washington, DC.

Metropolis, N., A.W. Rosenbluth, M.N. Rosenbluth, A.H. Teller, and E. Teller. 1953. Equation of state calculations by fast computing machines. *Journal of Chemical Physics* 21(6): 1087–1092.

Nataf, A. 1962. Determination des distribution dont les marges sont donnees. *Comptes Rendus de PAcademie des Sciences, Paris* 225: 42–43.

Ou, C.Y. and J.T. Liao. 1987. Geotechnical engineering research report, GT96008, National Taiwan University of Science and Technology, Taipei.

Phoon, K.K. and F.H. Kulhawy. 1999a. Characterizations of geotechnical variability. *Canadian Geotechnical Journal* 36(4): 612–624.

Phoon, K. K., Z. Cao, J. Ji, Y.F. Leung, S. Najjar, T. Shuku, C. Tang, Z.Y. Yin, I. Yoshida, and J. Ching. 2022. Geotechnical uncertainty, modeling, and decision making. *Soils and Foundations* 62(5): 101189.

Phoon, K.K. 2008. Numerical recipes for reliability analysis—A primer, Chapter 1. In *Reliability-Based Design in Geotechnical Engineering: Computations and Applications*. Taylor & Francis, New York, 1–75.

Phoon, K.K. and J. Ching. 2013. Multivariate model for soil parameters based on Johnson distributions. *Foundation Engineering in the Face of Uncertainty*, Geotechnical Special Publication honoring Professor F.H. Kulhawy, Reston, VA, 337–353.

Phoon, K.K. and F.H. Kulhawy. 1999b. Evaluation of geotechnical property variability. *Canadian Geotechnical Journal* 36(4), 625–639.

Phoon, K.K. 2020. The story of statistics in geotechnical engineering. *Georisk: Assessment and Management of Risk for Engineered Systems and Geohazards* 14(1): 3–25.

Rasmussen, C.E. and C.K.I. Williams. 2006. *Gaussian processes for machine learning*. MIT Press, London.

Rossi, R.J. 2018. *Mathematical statistics: An introduction to likelihood based inference*. John Wiley & Sons, New York.

Rubin, D.B. 1988. Using the SIR algorithm to simulate posterior distributions. *Bayesian Statistics* 3: 395–402.

Slifker, J.F. and S.S. Shapiro. 1980. The Johnson system: Selection and parameter estimation. *Technometrics* 22(2): 239–246.

Stein, M.L. 1999. *Interpolation of spatial data: Some theory for Kriging*. Springer, New York.

Stuedlein, A.W., S.L. Kramer, P. Arduino, and R.D. Holtz. 2012. Geotechnical characterization and random field modeling of desiccated clay. *Journal of Geotechnical and Geoenvironmental Engineering* 138(11): 1301–1313.

Vanmarcke, E.H. 1977. Probabilistic modeling of soil profiles. *Journal of the Geotechnical Engineering Division* 103(11): 1227–1246.

Vanmarcke, E.H. 1983. *Random fields – analysis and synthesis*. MIT Press, Cambridge, MA.

Yang, Z.Y. and J. Ching. 2021. Simulation of three-dimensional random field conditioning on incomplete site data. *Engineering Geology* 281: 105987.

Yoshida, I., Y. Tomizawa, and Y. Otake. 2021. Estimation of trend and random components of conditional random field using Gaussian process regression. *Computers and Geotechnics* 136: 104179.

Yuen, K.V. 2010. *Bayesian methods for structural dynamics and civil engineering*. John Wiley & Sons, New York.

Woodbury, M.A. 1950. *Inverting modified matrices*. Memorandum Rept. 42, Statistical Research Group, Princeton University, Princeton, NJ.

Index

Page numerals in *italics* refer to figures and pages in **bold** refer to tables

For Product Safety Concerns and Information please contact our EU representative GPSR@taylorandfrancis.com
Taylor & Francis Verlag GmbH, Kaufingerstraße 24, 80331 München, Germany

www.ingramcontent.com/pod-product-compliance
Lightning Source LLC
Chambersburg PA
CBHW060306220326
41598CB00027B/4250
9781032314433